Innovative
Applications of
Artificial Intelligence

Innovative Applications of Artificial Intelligence

Edited by Herbert Schorr and Alain Rappaport

AAAI Press / The MIT Press
Menlo Park • Cambridge • London

©1989 AAAI Press
445 Burgess Drive
Menlo Park, CA 94025

Copublished and distributed by The MIT Press, Massachusetts
Institute of Technology, Cambridge, Massachusetts, and London, England.

ISBN 0-262-69137-X (paperback); ISBN 0-262-19294-2 (casebound)

The chapter by Ted E. Senator, T. M. Witte, and Ted Kral, "Knowledge-
Based System Technology Applied to Naval Battle Managment Decision
Aiding" is approved for public release, distribution unlimited.

Objective-C™ is a registered trademark of the Stepstone Corporation.
PepPro™ and SpinPro™ rare registered trademarks of Beckman Instru-
ments, Inc. KEE™ is a registered trademark of IntelliCorp. Flavors™ and
MacIvory™ are registered trademarks of Symbolics, Inc. Packaging
Advisor™ is a registered trademark of the Du Pont Company. VAX™,
OPS5™, VAX BASIC™, and VMS™ are registered trademarks of Digital
Equipment Corporation. Underwriting Advisor™, Lending Advisor™, and
Syntel™ are registered trademarks of Syntelligence. ART™ and
Viewpoints™ are registered trademarks of Inference Corporation.
DesignAdvisor™ is a registered trademark of NCR Corporation. Golden
Common Lisp™ is a registered trademark of Gold Hill Computers, Inc.

Contents

Biotechology

The PepPro Peptide Synthesis Expert System
Matt Heffron, Philip R. Martz, Suresh Kalbag,

Emergency Services

A Knowledge-Based Emergency Control Advisory System

MannTall: A Rescue Operations Assistant

Law

An Expert System for Legal Consultation

Manufacturing Assembly

CAN BUILD: A State-of-the-Need Inventory Simulation Tool

Charley: An Expert System for
Diagnostics of Manufacturing Equipment

Harnessing Detailed Assembly Process Knowledge with CASE

Knowledge-Based Statistical Process Control
Kenneth R. Anderson, David E. Coleman, C. Ray Hill, Andrew P. Jaworski,

Logistics Management System

Manufacturing Design

Coolsys: A Cooling Systems Design Assistant
Patricia G. Friel, Richard J. Mayer, Jeffrey C. Lockledge,

Design Advisor: A Knowledge-Based Integrated Circuit Design Critic

OHCS: Hydraulic Circuit Design Assistant

Contributors

Kenneth R. Anderson, BBN Laboratories, 10 Moulton Street, Cambridge, Massachusetts 02138

John Aubrey, Australian Coal Industry Research Laboratories, Ltd. 14-30 Delhi Road, North Ryde, Sydney, Australia

Tomio Baba, Kayaba Industry Company, Ltd., Technical & Research Center, 1805-1 Asamizodai Sagamihara-Shi, Kanagawa 228, Japan

Paul Baffes, Artificial Intelligence Section, FM72, NASA Johnson Space Center, Houston, Texas 77058

Atul Bajpai, Advanced Engineering Staff, General Motors Corporation, GM Technical Center, Warren, Michigan 48090

Henry Brice, Ford Motor Company, 17000 Oakwood Boulevard, Allen Park, Michigan 48101

Martin Brooks, Center for Industrial Research, Box 124 Blindern, 0314 Oslo 3, Norway

Elizabeth Byrnes, Manufacturers Hanover Trust Company, 270 Park Avenue, New York, New York 10017

Thomas Campfield, Manufacturers Hanover Trust Company, 270 Park Avenue, New York, New York 10017

Ross S. Cann, Chase Lincoln First Bank, N.A., One Lincoln First Square, Rochester, New York 14643

Steinar Carlsen, Center for Industrial Research, Box 124 Blindern, 0314 Oslo 3, Norway

R. A. Chekaluk, GM Research Laboratories, General Motors Corporation, Warren, Michigan 48090

Tat-Leong Chew, Information Technology Institute, National Computer Board, 71, Science Park Drive, Singapore 0511, Republic of Singapore

Paul Clancy, Metropolitan Life Insurance Company, One Madison Avenue, New York, New York 10010

David E. Coleman, Alcoa Laboratories, Alcoa Center, Pennsylvania 15069

Bruce Connor, Manufacturers Hanover Trust Company, 270 Park Avenue, New York, New York 10017

Michael R. Craig, Chase Lincoln First Bank, N.A., One Lincoln First Square, Rochester, New York 14643

Douglas F. Dyckes, Department of Chemistry, University of Houston, Houston, Texas 77204

A. J. Finkel, Thomas J. Watson Research Center, International Business Machines Corporation, Yorktown Heights, New York 10598

Kenneth Fordyce, International Business Machines, Numerical and Technical Computing Center, Department 34EA, MS 284, Kingston, New York 12401

Patricia G. Friel, Knowledge-Based Systems Laboratory, Department of Industrial Engineering, Texas A&M University, College Station, Texas 77840

Andrew Gill, Information Technology Institute, National Computer Board, 71 Science Park Drive, Singapore 0511, Republic of Singapore

Georges Girod, Mediatop, 81 bd Berthier, 75017 Paris, France

Peter E. Hart, Syntelligence, 1000 Hamlin Court, Sunnyvale, California 94089

Scott Hatfield, Ford Motor Company, 17000 Oakwood Boulevard, Allen Park, Michigan 48101

Matt Heffron, Beckman Instruments, Inc., 2500 North Harbor Boulevard, Fullerton, California 92634

Troy A. Heindel, National Aeronautics and Space Administration, Lyndon B. Johnson Space Center, Houston, Texas 77058

C. Ray Hill, Alcoa Laboratories, Alcoa Center, Pennsylvania 15069

Gerald Hoenig, Metropolitan Life Insurance Company, One Madison Avenue, New York, New York 10010

Atle Honne, Center for Industrial Research, Box 124 Blindern, 0314 Oslo 3, Norway

Grace Hua, Computer Sciences Corporation, 16511 Space Center Boulevard, Houston, Texas 77058

E. M. Hufziger, GM Research Laboratories, General Motors Corporation, Warren, Michigan 40890

Michael A. Hutson, General Motors Acceptance Corporation, Forward Systems, 3044 West Grand Boulevard, Detroit, Michigan 48202

Andrew P. Jaworski, Alcoa Laboratories, Alcoa Center, Pennsylvania 15069

Wayne P. Johnson, Inference Corporation, 5300 West Century Boulevard, Los Angeles, California 90045

Suresh Kalbag, Genetech, Inc., 460 Point San Bruno Boulevard, South San Francisco, California 94080

Kyle W. Kindle, Chase Lincoln First Bank, N.A., One Lincoln First Square, Rochester, New York 14643

Kirsten Y. Kissmeyer, The MITRE Corporation, Burlington Road, Bedford, Massachusetts 01730

Ted Kral, Defense Advanced Research Projects Agency (DARPA), 1400 Wilson Boulevard, Arlington, Virginia 22209

Robin M. Krumholz, Digital Equipment Corporation, 5 Carlisle Road, Westford, Massachusetts 01886

Patrick Landry, Intellia, 11 rue Mercoeur, 75011 Paris, France

Joo-Hong Lim, Information Technology Institute, National Computer Board, 71 Science Park Drive, Singapore 0511, Republic of Singapore

Jeffery C. Lockledge, Knowledge-Based Systems Laboratory, Department of Industrial Engineering, Texas A&M University, College Station, Texas 77840

R. Bowen Loftin, University of Houston-Downtown, One Main Street, Houston, Texas 77002

Patrick L. Love, Alcoa Laboratories, Alcoa Center, Pennsylvania 15069

Richard Marczewski, Advanced Engineering Staff, General Motors Corporation, GM Technical Center, Warren, Michigan 48090

Thomas J. Martin, Arthur D. Little, Inc., 35 Acorn Park, Cambridge, Massachusetts 02140

Philip R. Martz, Beckman Instruments, Inc., 2500 North Harbor Boulevard, Fullerton, California 92634

Richard J. Mayer, Knowledge Based Systems Laboratory, Department of Industrial Engineering, Texas A&M University, College Station, Texas 77840

William J. McClay, Boeing Computer Services, Electronics Support, PO Box 24346, MS 7L-45, Seattle, Washington 98124

Robert Z. McFarland, Unisys Corporation, P.O. Box 500, Blue Bell, Pennsylvania 19424

K. R. Milliken, Thomas J. Watson Research Center, International Business Machines Corporation, Yorktown Heights, New York 10598

John F. Muratore, National Aeronautics and Space Administration, Lyndon B. Johnson Space Center, Houston, Texas 77058

Terri B. Murphy, National Aeronautics and Space Administration, Lyndon B. Johnson Space Center, Houston, Texas 77058

Yusei Nakashima, Kayaba Industry Company, Ltd., Technical & Research Center, 1805-1 Asamizodai Sagamihara-Shi, Kanagawa 228, Japan

Zoltan Nemes-Nemeth, Australian Coal Industry Research Laboratories, Ltd., 14-30 Delhi Road, North Ryde, Sydney, Australia

John O'Brien, Ford Motor Company, 17000 Oakwood Boulevard, Allen Park, Michigan 48101

Patrice Orgeas, Mediatop, 81 bd Berthier, 75017 Paris, France

Alain Rappaport, Neuron Data, 444 High Street, Palo Alto, California 94301

Arthur N. Rasmussen, The MITRE Corporation, Burlington Road, Bedford, Massachusetts 01730

Douglas K. Reece, E. I. du Pont de Nemours & Company, Inc., Wilmington, Delaware 19898

Scott A. Richardson, NCR Microelectronics, 2001 Danfield Court, Fort Collins, Colorado 80525

R. Douglas Riecken, 640 Sherman Avenue, Plainfield, New Jersey 07060

Kenin Sahin, Consultants for Management Decisions, Inc., One Broadway, Cambridge, Massachusetts 02142

Keith Sawyer, Consultants for Management Decisions, Inc., One Broadway, Cambridge, Massachusetts 02142

Arnold Schmitt, Metropolitan Life Insurance Company, One Madison Avenue, New York, New York 10010

Herbert Schorr, USC Information Sciences Institute, 4676 Admiralty Way, Marina del Ray, California 90292

Ted. E. Senator, Department of the Navy, Information Resources Management, Washington, DC 20350

Roger C. Shulze, Chrysler Corporation, 12000 Chrysler Drive, Highland Park, Michigan 48203

Marwan Simaan, Electrical Engineering Department, Benedum Hall, University of Pittsburgh, Pittsburgh, Pennsylvania 15261

Gary M. Smith, Chrysler Corporation, 12000 Chrysler Drive, Highland Park, Michigan 48203

Douglas A. Spindler, Alcoa Laboratories, Alcoa Center, Pennsylvania 15069

Robin L. Steele, NCR Microelectronics, 2001 Danfield Court, Fort Collins, Colorado 80525

Gerald Sullivan, International Business Machines, Advanced Industrial Engineering, Department 746, Building 965-3, Essex Junction, Vermont 05455

Anne M. Tallant, The MITRE Corporation, Burlington Road, Bedford, Massachusetts 01730

John A. Thompson, Boeing Computer Services, Electronics Support, PO Box 24346, MS 7L-45, Seattle, Washington 98124

Alvin S. Topolski, E. I. du Pont de Nemours & Company, Inc., Wilmington, Delaware 19898

Paul Voelker, Beckman Instruments, Inc., 2500 North Harbor Boulevard, Fullerton, California 92634

Norman B. Waite, Thomas J. Watson Research Center, International Business Machines Corporation, Yorktown Heights, New York 10598

Lui Wang, Artificial Intelligence Section, FM72, NASA Johnson Space Center, Houston, Texas 77058

Michael A. Winchell, NCR Microelectronics, 2001 Danfield Court, Fort Collins, Colorado 80525

T. M. Witte, Space and Naval Warfare Systems Command (SPAWAR), Washington, DC 20363

Richard Woodhead, Inference Corporation, 5300 West Century Boulevard, Los Angeles, California 90045

Preface

Herbert Schorr

Scientific achievements from the field of artificial intelligence have enabled us to tackle new problems in which the computer formerly played no role. While we have not yet achieved all the goals of automating cognition, years of research are nevertheless being rewarded by having a major impact on everyday operations in the workplace.

We hope this book illustrates how the implications of AI research and development are beginning to affect all areas of today's world, particularly those areas that concern commerce and industry. This collection of articles discusses what really works and what the problems are as AI makes the transition from the research laboratory to industrial practice. Tackling these new and innovative applications will lead to better technology because we will find and remedy current deficiencies when solving real problems.

At the current rate of AI's development, the definition of innovation itself changes rapidly. Some innovative applications advance the state of the underlying AI technology. Others involve the integration of AI with standard data-processing systems. Some applications are likely to be the last of a first generation whose developments were heroic acts. Most of the examples of innovation treated here, however, illustrate how the use of AI technology has become not only effective but also economically feasible. They show the integration of AI into the standard data-processing environments as well as its use in the softer sciences and in the management of human affairs.

The applications selected for inclusion here (chosen to illustrate the field's breadth rather than exhaust any one domain) show AI to be a successful tool in a wide spectrum of domains and tasks. Nearly all are expert systems because it is in this form that AI is most rapidly coming into widespread use. Why is this the case? To find an answer, let us look at robotics and manufacturing, neural networks, and natural-language processing.

Robotics can be defined as the ability to have humanlike machines with arms, hands, eyes, and legs, or, alternatively, the creation of intelligent machines with computer control and sensing. Robots have been used for some time in such applications as paint spraying, welding, inspection, pick-and-place operations, electrical component or board insertion, and aircraft assembly. Innovative applications have not come about. Robots remain difficult to program, and they suffer from a lack of robustness. A good robot software system must have real -time capability. It must also provide for multiple subsystems operating in parallel, handle changing environments, integrate often unreliable sensors, control tightly coupled electrical and mechanical subsystems, obtain complex goals from a human operator, and convert high level goals into network commands. So far, no such robot software system for complex tasks is commercially available. By contrast, large industrial strength expert-system shells are readily available. Hence, robots seem to be stuck with their early applications and have made small commercial progress in the last few years.

Neural networks have recently attracted a large amount of attention. In terms of maturation, this area is in the embryonic state expert systems occupied in the early 1980s. Before neural network concepts can be applied to any problem, many issues have to be resolved. For example, input and output representations must be chosen, and the number of layers, the interconnections, and the methods of training must all be understood. A large collection of test cases with correct responses needs to be gathered. At the time of this writing, we know of no neural networks in practical day-to-day use, and the present collection does not report any neural network application Thus, while this technology appears to possess vast potential in financial, manufacturing, defense, and other areas, we leave it for this book's successor to cover such applications.

Natural language processing has been constrained historically by limitations of computational power, but the fantastic progression of computational cost/performance has eliminated this bottleneck. Natural language products are therefore available, but most employ mainly syntactic processing, with some semantics, and very little pragmatics—a major problem. Domain knowledge and lexicon must still be input manually by the user. Thus today's applications — database interfacing, text processing, and some machine translation — are very limited and very few low-level natural language functions are being deployed. This book does describe one new application, the translation of interbank money transfer telexes. Most of the other natural language applications, predominately in database query, have been around for some time.

So let us return to expert systems. Why is this field of AI taking off? One reason involves a change in basic impulse. Until recently, the field has been technology driven. The expert system shell technology was developed so that it was reusable. Next, the domains of the technology's applicability began to be understood, so that expert systems could be developed with reasonable expenditures of manpower. Applications began to be deployed that promised strategic corporate advantage. Today, the market is beginning to be market driven. As the reader will be reminded here, there are now many application experiences to guide product needs and which have demonstrated that expert systems derive a substantial return on investment (ROI).

Expert systems research has therefore moved into what we believe is the third stage of growth. The first stage was the demonstration of the technology's feasibility by the universities. Complex domains, such as the diagnosis of illnesses, language translation, and geology were studied. During the transition to the commercial world, the second stage, the technology was used to do such mundane tasks such as claim settlements, credit evaluations, testing, and training. The main users of these expert systems tended not to be experts of the highest caliber, but rather clerks and technicians whose capabilities the AI products sought to supplement. However, if expert systems were to be useful to nonexperts in this way, the applications could no longer stand alone. Used by a clerk, for example, an expert system doing credit evaluations needed to access data containing the financial history of a company or individual. Expert systems useful to the oil industry required access to existing Fortran programs that did seismic analysis. Driven by such necessities, we now enter the field's third developmental state, illustrated in various chapters of this book, where knowledge-base systems are being integrated with both database systems and conventional procedural programs on conventional platforms. Making this integration transparent, enabling the result to perform well, and refining applications that truly complement corporate databases will provide the discipline's primary challenges over the next few years.

In the marketplace customers continue to buy. Although there has been some talk of an AI winter, the field's compound growth, as reported by many independent surveys, continues at a healthy rate—growth coming not from R & D activity, but from development and deployment of production applications. To date we have produced very few strategic lines of business applications, but they are coming. Applications for the manufacturing, finance, insurance, and process industries are being developed most actively, as can be seen in the appropriate following sections of this book. Customer demand is also strong for cross-industry applications. Such applications include systems management, (see the Expert Operator chapter herein), office systems, and database management.

Expert systems are solving problems today. They will continue to have a profound effect on computer systems and their architecture; on programming languages, databases, and user interfaces; and most importantly on the application development environment. Neural nets, we suspect, are likely to be the next technology to become commercially useful because they, like expert systems, hold the promise of providing solutions to strategic and previously unsolved problems. Very complex robotics, machine-based natural language processing, machine learning, and speech recognition will follow sometime in the future. AI is a very fertile field from which many new important applications will grow. This book illustrates just the beginning of the transition of AI to the commercial world.

Aerospace

Space Shuttle Telemetry Monitoring by
Expert Systems in Mission Control
NASA, The MITRE Company, and Unisys Corporation

An Intelligent Training System
for Space Shuttle Flight Controllers
NASA, University of Houston, Computer Sciences Corporation

Space Shuttle Telemetry Monitoring by Expert Systems in Mission Control

*John F. Muratore, Troy A. Heindel, Terri B. Murphy,
Arthur N. Rasmussen, and Robert Z. McFarland*

The successful launch of the space shuttle *Discovery* in September 1988 represented a bright new beginning for NASA. It also represented a new start for the Mission Control Center (MCC) at the Lyndon B. Johnson Space Center. For the first time, knowledge-based systems were used at this critical facility, which is the focal point for space shuttle flight operations. Knowledge-based systems were used to monitor the space shuttle, detect faults, and advise flight operations personnel. This application was the first time knowledge-based system technology had been used in a NASA spaceflight operational environment. Flight-management decisions were made directly based on the results of knowledge-based systems. This achievement marks a milestone in the application of AI.

In the past, MCC has relied on mainframe-based processing and display techniques, which emphasized the use of highly skilled personnel, known as flight controllers, to monitor data, detect failures, analyze system performance, and make changes to flight plans. Figure 1 illustrates the working environment of MCC and figure 2 shows an example of the displays provided by this mainframe-based system. This setup

Figure 1. Space Shuttle Mission Control. Courtesy NASA.

worked well for programs of short duration, but as NASA looked to operating long-duration programs, such as the space station *Freedom*, flight operations personnel began to explore techniques to automate mission monitoring, display and analysis functions.

During the recovery from the *Challenger* accident, two real-time expert systems were implemented and certified for use in making flight-critical decisions. This work was performed using commercial hardware and software by flight controllers and knowledge engineers from a combined NASA- industry team. These two expert systems, which monitor space shuttle communications and the space shuttle main engines, were successfully used in the STS-26 *Discovery* flight.

The Problem

The MCC information systems are vital to the safety and success of manned spaceflights conducted by the United States. The centralized system currently employed presents only raw data to flight controllers, with little interpretation. All processing is contained in a single, large, mainframe computer with software that is difficult to change and verify. This system presents data to the flight controllers but not informa-

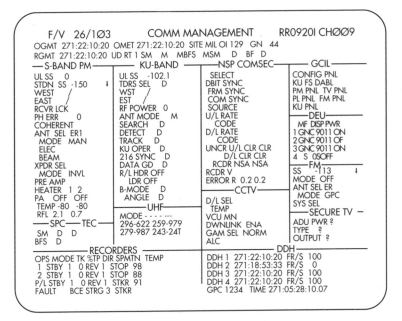

Figure 2. Typical MCC Mainframe Display.

tion. It is the job of the flight controllers to convert these raw data to information that can be used to manage the mission. Teaching flight controllers to perform this task is a major training problem that typically takes two to three years to complete. The nature of the mainframe-based MCC system, which largely provides monochrome text displays, causes even the most highly trained and motivated flight controllers to make occasional monitoring errors. In the MCC problem domain, a flight controller error can result in grave consequences.

The problem of ensuring high-quality decisions by the flight control team using this minimal information-processing capability is further complicated by NASA's unique bimodal age distribution. Because of the hiring freezes between the Apollo and space shuttle programs, the majority of NASA personnel fall into two distinct groups: the younger "shuttle-only" group under 35 years and the more experienced "Apollo-era veterans" of greater than 45 years. These veterans represent a dwindling supply of corporate knowledge and experience as they are promoted, retire, or move on to other activities.

As NASA moves into the space station era, it is further confronted by the requirement to continuously operate the station for its 20-year lifetime. It is unreasonable to expect that we will be able to maintain a large work force of highly skilled flight controllers to perform the demanding work loads which are imposed by our current style of system monitoring for this amount of time.

NASA'S Solution

AI is a natural source of techniques for converting data into information, capturing corporate knowledge, and lowering operator training and response time. In 1987, Mission Operations Directorate at the Lyndon B. Johnson Space Center started a project to apply the techniques and methodologies of AI, such as expert systems and natural language interfaces, to real space mission operations problems.

The first task of this project was to provide an "intelligent associate" to the flight controller monitoring the space shuttle's communications and data systems. This associate expert system, named for the first flight controller position selected for automation, is called the integrated communications officer (INCO) expert system. Development was started in August 1987, and the system was placed in MCC in April 1988. Approximately eight person-years worth of effort and $400,000 of hardware were required to field this system. The system was placed next to the INCO console, allowing operators to compare results of the conventional console to those of the expert system (figure 3). The INCO system was used during the STS-26 flight of *Discovery* in September 1988.

The INCO expert system was implemented on a conventional color graphics engineering workstation. The Masscomp 5600 with the Unix operating system was chosen because it is compatible with other workstations.

Automated monitoring was performed using both algorithmic and heuristic techniques. Knowledge was represented procedurally in C language code and rules. The representation for specific knowledge was driven by the complexity of the knowledge and the required rate of execution. In the Masscomp environment, conventional C programming generally executes faster than the interpreter in a rule-based system.

The procedural representations were built in a structured natural language and translated into C. The translation was done by a tool created by the project called computation development environment (CODE). CODE allows flight controllers to specify monitoring algorithms in a high-level language and then generates the C code necessary to perform the algorithm. The rule-based representation for both algorithms and heuristics was built using the C language inference production system (CLIPS) expert system tool developed by the Mission Planning and Analysis Division at the Lyndon B. Johnson Space Center.

One requirement placed on the INCO expert system was that it ad-

*Figure 3. Expert System Workstation Installed Next to
Conventional Console in MCC During STS–26. Courtesy NASA.*

vise flight controllers in real time. This requirement typically means failure detection within five seconds of an event, which requires the expert system to have direct electronic access to the real-time telemetry from the space shuttle. Another requirement for the system was that it be isolated from all the existing MCC systems so that problems in the stand-alone expert system did not affect flight-critical mainframe processing. The combination of these two requirements forced us to build a completely independent real-time telemetry-processing capability into the INCO expert system.

The stringent demands of executing processes supporting real-time telemetry processing while operating under an unmodified Unix led to the development of an innovative architecture for meeting real-time, knowledge-based system needs. The architecture was based on a four-layer model (figure 4): Raw data enter the first layer; as they move up through the layers, they are converted to higher-quality information.

The first layer performs basic data-acquisition tasks such as telemetry decommutation. This layer is performed by a commercial telemetry hardware device that transfers data into the Masscomp by way of a direct memory access interface.

The second layer contains generic data-conversion algorithms that do not require domain-specific knowledge. For example, algorithms that convert telemetry data between different floating-point formats are contained in this layer. This layer is performed in C on the Masscomp workstation.

The third layer uses procedural techniques to implement domain-specific knowledge. This knowledge is entirely algorithmic in nature. For example, in this layer, the system can monitor a voltage and signal an alarm if the voltage is below a required level. These algorithms were built using CODE.

The fourth layer uses rule-based techniques to represent both algorithmic and heuristic knowledge. It is often easier to implement complex algorithms, such as those which perform overall systems analysis, in the rule base rather than in the layer three procedures. The desire to use rule-based techniques must be balanced with concerns about the speed of execution. Items that had to execute once every second were implemented using the layer three procedural techniques.

Rules execute as an embedded component of the entire system. Rule-based components are only called into operation when the failure-detection algorithms at the third layer notice a significant change in the system status. In this way, we improve overall system performance. The rules are implemented in CLIPS and communicate with layer three procedures by way of a shared memory.

An interesting characteristic of this layered approach is that as data move up the layers, the total amount of data decreases, but the information value of the data increases. For example, 192,000 bits of information enter layer one, but the rule-based expert system only operates on 350 facts generated by the layer three algorithms. These 350 facts contain important verified information about the system, whereas alone the raw telemetry bits do not uniquely identify conditions on the spacecraft.

Failures are detected by the system and flagged to the operator in less than five seconds on a color graphics interface. Figure 5 shows a typical display. This figure contains all the information from the mainframe system display shown in figure 2.

On several occasions during ground simulations and shuttle flights, the system detected failures that were undetected by the flight control team. Sufficient confidence in the system has been gained so that conventional equipment is beginning to be replaced by expert systems, and some staffing reductions will occur.

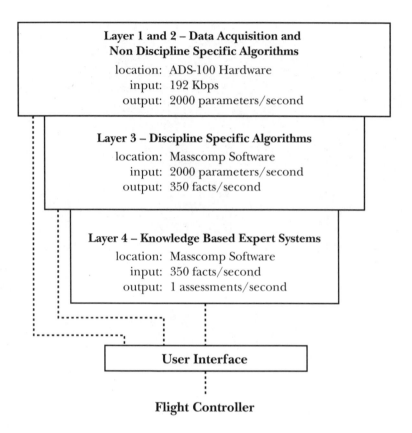

Figure 4. Layered Architecture of INCO Expert System.

Experience and Payoff

When work was started on this expert system, considerable debate went on about the expected payoff. Viewpoints centered around three areas. Each of these three views required different emphasis during expert system development.

The first view was to increase the quality and productivity of our current experienced personnel. Proponents of this viewpoint believed that the expert system should be developed to allow an experienced operator to make better flight decisions. This goal could be achieved by developing a system capable of continuously analyzing all incoming telemetry to a depth that would be impractical even for an expert operator.

The potential dollar value of this payoff is difficult to quantify. A single, good decision that allowed the completion of a several hundred

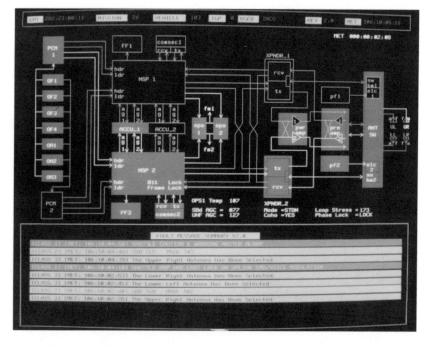

Figure 5. INCO Expert System Schematic Display. Courtesy NASA.

million dollar mission would clearly pay for a large amount of expert system work, but it would be difficult to say the good decision was completely the result of the expert system. Developing an expert system to meet this goal requires the incorporation of deep knowledge about a given spacecraft monitoring task.

The second view was to use expert systems to allow less experienced personnel to perform at the level of more senior personnel. This view has a measurable cost benefit in that it allows shorter training times. This viewpoint requires an emphasis on breadth of knowledge as well as improved human-computer interfaces.

The third view was that the use of expert systems should allow true staffing reductions by automating part of the systems monitoring job. This view requires deep knowledge about specific tasks as well as concentration on fault annunciation software and reliability issues.

In Mission Control, we have taken the approach that our first priority is the quality of decisions. The INCO expert system was initially designed to allow an expert INCO to function more productively by relieving the person of the mechanical tasks associated with scanning data. Our second priority was to present data to operators so that less experienced persons could operate at high proficiency levels. Our lowest priority was to reduce personnel. The INCO expert system does meet all these goals to a varying extent.

The INCO expert system captured deep knowledge about monitoring the space shuttle and, in fact, does allow operators to perform a more thorough job of monitoring telemetry. It contains knowledge about the monitoring task that is probably only shared by five or six experts, allowing even experienced INCOs to operate with the benefit of this expertise.

Based on the experience with INCO, we feel it is possible to reduce the training time for a first-time flight controller from 2 years to approximately 1.5 years. Four operators are in training at any time in the INCO area, resulting in a potential payoff of approximately one person-year per year (approximately $80,000 per year).

Unfortunately, we will not be able to achieve this benefit as long as it is required to train the flight controllers in both the conventional mainframe system and the expert system. We are trying to increase the reliability and operator confidence in this system to allow for total commitment to the expert system approach. In support of this goal, we removed two mainframe monitors from the conventional INCO consoles and replaced them with display units from expert system workstations. These workstation displays were used as primary tools by the INCOs for the STS-29 mission in March 1989. After this mission we removed two more mainframe display units and used them as primary tools in STS-30. We expect to realize some training benefits in early 1990.

Because of the INCO expert system, in late 1989 we will be able to reduce the size of the INCO monitoring team from four operators per shift to three. Because we currently have five teams of operators, this reduction will allow us to save approximately five person-years per year (approximately $400,000 per year).

It is important to realize that this payoff will not result in lower operations costs or lower levels of staffing. The space shuttle program is under constant pressure to fly more often and to exact more performance from every flight. This achievement requires additional staffing. The personnel reduction from the expert system will be reinvested to form a sixth INCO flight control team. This new team will allow us to meet the demands of the 1990 space shuttle flight schedule. Use of the expert system has made a new source of available trained flight controllers, which we can apply to new problems and higher flight rates. The system has allowed us to do more for the same money rather than do the same for less money.

The INCO expert system hardware investment was $400,000 and approximately eight person-years of effort ($640,000) over two years. With the reduction of the INCO team scheduled for late 1989, we will achieve a cost payback in approximately two years. The cost payback, however, is not nearly as important as the fact that the system will allow us to better use our precious and scarce resource of experienced personnel.

Operator Acceptance

The real measure of success for this enterprise is the acceptance of real-time expert system technology by the mission operation community. Because the system was developed primarily by flight controllers, acceptance in mission operations occurred almost immediately. A rapid prototyping methodology that allowed us to react rapidly to changes suggested by the flight control team also increased acceptance. On several occasions when the mainframe complex failed during simulations, flight controllers never hesitated to use the expert system as their only basis for flight decisions.

The degree of acceptance was dramatically demonstrated by the chain of events that led to our second expert system. In May 1988, the shuttle flight controllers responsible for monitoring the main engines determined there were several failure modes of the main engines that required automated fault detection. Flight controllers could not manually perform calculations and assessments fast enough to meet the demands of monitoring this high-performance system in dynamic flight.

The necessary fault-detection routines were designed and built using the first three layers of the INCO expert system. All the knowledge in this main engine system was algorithmic in nature and of low complexity. The nature of the knowledge combined with the high-speed requirements for decisions during ascent led to the decision to place all our efforts in the first three layers.

Development started in May 1988, and the system was certified for use in August 1988. Three full-time staff members were assigned to this task, which was called the booster expert system. The booster expert system provided a new and significant capability to Mission Control. Booster was certified for use in making flight-critical decisions and was used during the STS-26 flight. Booster is a major payoff to NASA because it has greatly improved the quality of flight decisions during the dynamic ascent phase.

Future Efforts

Based on the STS-26 experience, this effort is being expanded into multiple new disciplines such as mechanical systems; electric power;

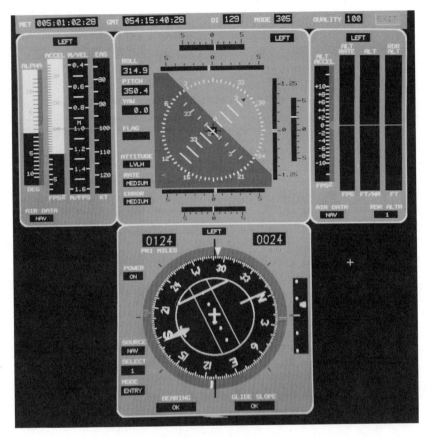

*Figure 6. Workstation Simulation of
Shuttle Flight Instruments. Courtesy NASA.*

and guidance, navigation, and control. In at least two areas, the successful implementation of the expert system will result in small staffing reductions. Each of these new systems represents new monitoring challenges, but the basic layered architecture of the INCO expert system will be used.

The general applicability of this architecture was proved in 1988 when the system was used by NASA and the United States Air Force (USAF) for monitoring telemetry from experimental aircraft at the NASA Dryden Flight Research Facility and the Air Force Flight Test Center at Edwards Air Force Base in California. In each of these cases, telemetry data from an experimental aircraft were incorporated into the INCO structure. In both cases, the system was modified to the aeronautics applications in less than 48 hours. Use of this structure will allow operations personnel at these facilities to concentrate on expert system knowledge base issues rather than on the development of another real-time environment.

The different expert systems will be connected by an Ethernet in late 1989. This move will allow us to experiment with cooperation between multiple expert systems. Just as the various flight controllers communicate and work together in a control room, the local area network will allow the expert systems to work together.

Part of the human-computer interface for the guidance, navigation and control expert system will include graphic displays of the astronaut's flight instruments on ground workstations by reconstructing the displays from telemetry. The attitude direction indicator (ADI), or astronaut's artificial horizon instrument, emulation is already complete, and was used during STS-29 and STS-30 missions.

The INCO expert system project is the first significant operational use of knowledge-based system technology in a NASA operational environment. It has shown that expert systems can play an integral role in manned spaceflight operations. As NASA moves forward in its future space activities, expert systems will be there.

Acknowledgments

The INCO expert system project was cofunded by three NASA organizations: the Office of Aeronautics and Space Technology, the Office of Space Station, and the Office of Space Flight. The effort also received funding from the joint USAF-NASA Advanced Launch System Project Office. The effort was developed by a joint NASA-industry team, including members from Rockwell Space Operations, Mitre Corporation, Unisys, and Dual and Associates. The authors wish to thank each of these organizations for their support. The authors wish to acknowledge the programming support provided to INCO by Thomas Kalvelage, Daryl Brown, Glenn Binkley, Erick Kindred, Cheryl Whittaker, and Debbie Horton. The booster expert system was designed and programmed by Michael Dingler. The ADI emulation was designed and programmed by Mark Gnabasik. Installation support was provided by James Gentry of the Bendix Field Engineering Corporation. John Bull and Peter Friedland, from Ames Research Center, and Melvin Montemerlo, Lee Holcomb, Gregg Sweitek, and Charles Holliman, NASA Headquarters, provided valuable comments and insights throughout the development of the project. Robin Madison at the US Airforce Flight Test Center and Dale Mackall at Dryden Flight Research Center have modified our work for aeronautic applications.

An Intelligent Training System for Space Shuttle Flight Controllers

R. Bowen Loftin, Lui Wang, Paul Baffes, and Grace Hua

The Mission Operations Directorate (MOD) at National Aeronautics and Space Administration Lyndon B. Johnson Space Center (JSC) is responsible for the ground control of all Space Shuttle operations. Those operations that involve alterations in the characteristics of the space shuttle's orbit are guided by a flight controller, the flight dynamics officer (FDO), sitting at a console in the front room of the Mission Control Center (MCC). Currently, the training of the FDOs in-flight operations is principally accomplished through the study of flight rules, training manuals, and on-the-job training in integrated simulations. Two to four years are normally required for a trainee FDO to be certified for many of the tasks that must be performed during space shuttle missions. On-the-job training is highly labor intensive and presupposes the availability of experienced personnel with both the time and the ability to train novices. As the number of experienced FDOs decreases through retirement, transfer (especially of United States Air Force personnel), and promotion and as the preparation for, and actual control of, missions occupies most of the available MCC schedule, on-the-job training has become increasingly difficult to provide to novice FDOs.

As a supplement to the existing modes of training, the Artificial Intelligence Section (AIS) at the space center has developed an autonomous, intelligent, computer-aided training system. The system trains inexperienced flight controllers in the deployment of a payload-assist module (PAM) satellite from the space shuttle. This task is complex and mission-critical; it requires skills used by the experienced FDO in performing many of the other operations that are the FDO's responsibility.

Description of the Application

Since the first proposals to apply AI to the tutoring or training task (Carbonell 1970, Hartley and Sleeman 1973), a large number of systems have been developed for academic and industry-government training environments (Sleeman and Brown 1982, Yazdani 1986, Kearsley 1987, Wenger 1987). In spite of these efforts, few completed systems have come into widespread use for routine training.

The training system is designed to aid novice FDOs in acquiring the experience necessary to carry out a PAM deployment in an integrated simulation. It is intended to permit extensive practice with both nominal deployment exercises and others containing typical problems. After successfully completing training exercises that contain the most difficult problems, together with realistic time constraints and distractions, the trainee should be able to successfully complete an integrated simulation of a PAM deployment without aid from an experienced FDO. The philosophy of the payload-assist module deploys/intelligent computer-aided training (PD/ICAT) system is to emulate, to the extent possible, the behavior of an experienced FDO devoting full time and attention to the training of a novice. Such training would include proposing challenging training scenarios, monitoring and evaluating the actions of the trainee, providing meaningful comments in response to trainee errors, responding (if appropriate) to trainee requests for information and hints, and remembering the strengths and weaknesses displayed by the trainee so that appropriate future exercises can be designed.

The PD/ICAT system architecture consists of five components and is organized around a common blackboard to facilitate communication among the different components (Loftin et al. 1988).

User Interface

The user interface permits the trainee to access the same information that would be available in MCC. The interface also serves as a means for the trainee to take actions and receive feedback from the training

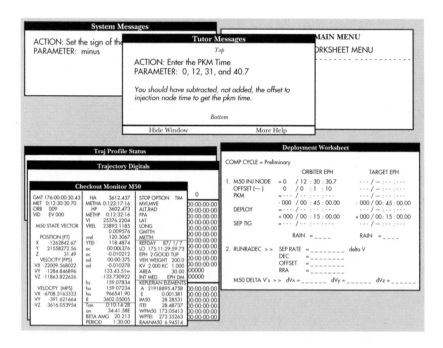

Figure 1. A Typical Screen as Seen by a User of the PD/ICAT System.
Menus at the upper right allow the trainee to interact with the system.
Displays, identical to those used in MCC, appear in the lower left.
A worksheet used by FDO in completing the task is shown in the lower
right. System messages are displayed in the upper left, and error-help
messages are provided in a pop-up window near the center of the screen.

session manager. To the degree possible on a single screen, the interface mimics the environment of the domain in which the trainee will eventually work. Figure 1 shows a typical screen display.

Domain Expert

A domain expert, referred to as the deploy expert (DeplEx), is in the form of a production-rule system capable of carrying out the satellite deployment process using the same information available to the trainee. DeplEx also contains a list of *mal-rules* (explicitly identified errors that novice trainees commonly make [Sleeman and Brown, 1982]) so that the trainee can be provided with feedback specifically designed to help the individual overcome any anticipated conceptual or procedural problems.

Training Session Manager

The training session manager (TSM) consists of two expert systems: (1) an error-detection component that compares the assertions made by DeplEx (of both correct and incorrect actions in a particular context) with those made by the trainee and (2) an error-handling component which decides on the appropriate method of guidance based on the trainee's skill level.

Trainee Model

The trainee model is an object-oriented data structure that contains a history of the individual trainee's interactions together with summary evaluative data. The model also has a report-generation feature that produces a formatted trace of each trainee session and provides the trainee's supervisor with a high-level description of the trainee's current skill level and progress.

Training Scenario Generator

The training scenario generator (TSG) is an expert system and database that designs increasingly complex training exercises based on the current skill level contained in the trainee's model and any weaknesses or deficiencies that the trainee exhibited in previous interactions. The database serves as a repository for all parameters needed to define a training scenario and includes problems or abnormalities of graded difficulty. The nature of this component of the PD/ICAT architecture is fully developed elsewhere (Loftin, Wang, and Baffes 1988).

System Integration

All the expert system components of PD/ICAT (DeplEx, TSM, and TSG) communicate by way of a common blackboard. The blackboard also contains a representation of the current trainee action(s) and is a source of data for updating the trainee model.

Innovative Features of the Application

Although a number of intelligent training systems have been developed (for example, sophisticated instructional environment (SOPHIE) [Brown, Burton, and de Kleer 1982] and Steamer [Hollan, Hutchins, and Weitzman 1984]), few, if any, can be said to have been deployed. The PD/ICAT system was deployed and is in use for training novice flight controllers and providing experienced personnel with practice and refresher training.

Many features of the PD/ICAT system are innovative in their own right or are innovative in their application to intelligent training systems.

PD/ICAT is composed, in part, of four expert systems that cooperate through and communicate by means of a common blackboard. This approach was used to permit the segregation of domain-independent knowledge so that the system architecture could be adapted to different training tasks.

Unlike most intelligent tutoring-training systems, PD/ICAT does not require the trainee to follow a single, correct path to the solution of a problem. Rather, a trainee is permitted to select any correct path, as determined by the scenario context. The method used to accomplish this flexibility, without generating a combinatorial explosion of solution paths, is believed to be unique.

Error detection occurs through the comparison of the trainee's actions with those of an expert. In the case of complex actions, the error detection is made at the highest level to avoid confusing the trainee by detecting all errors that propagated from the one deemed most significant.

Error handling can be accomplished through the matching of trainee actions with mal-rules containing errors that are commonly made by novices. In addition, based on the trainee model, the TSM error-handling component decides what type of feedback to give the trainee. Explanations or hints can be detailed for novices and quite terse for experienced personnel. In some cases, TSM might decide not to call attention to the error if a reasonable probability exists that the trainee might catch the mistake.

TSG examines the trainee model and creates a unique scenario for each trainee whenever a new session begins. This scenario is built from a database containing a range of typical parameters describing the training context as well as problems of graded difficulty. Scenarios evolve to greater difficulty as the trainee demonstrates the acquisition of greater skills in solving the training problems.

At the conclusion of each session, the trainee is provided with a formatted trace of the session that highlights the correct and incorrect actions taken, the time required to complete the exercise, and the type of assistance provided by the system. In addition, the trainee's supervisor can view a global history of each trainee's interaction with the system and even generate graphs of trainee performance measured against a number of variables.

Criteria for a Successfully Deployed Application

From the inception of the PD/ICAT project, a number of factors were recognized as essential for success at meeting objectives and gaining acceptance by the intended audience. First, the involvement of the ultimate users (or customers) throughout the development process was certainly the most important factor in PD/ICAT's successful deployment. This involvement allowed the development team to have at hand the experts in the targeted procedure and to quickly test competing approaches to the solution of specific development problems.

Second, the intended audience for the training system was asked to provide commentary at each stage of the interface development. Thus, at the project's conclusion, the interface had already achieved tacit acceptance by those persons it was intended to serve.

Third, built-in maintainability is provided by the use of production rules in those components of PD/ICAT that would require alteration as the procedures it was designed to teach are altered. After fielding a number of production-rule systems, AIS found, that such systems can easily be altered and reverified as the target tasks change.

Fourth, sufficient training in production-rule coding and detailed documentation were provided to system users so that they were able to provide their own long-term support for PD/ICAT.

Fifth, although PD/ICAT was developed on a Symbolics Lisp machine using ART and Lisp, it was ported to a Unix-based workstation using C and C language integrated production system (CLIPS) (a production-rule system written in C and developed by AIS at the space center). Such workstations were already in use by PD/ICAT's intended users; thus, they were not required to purchase and learn a different hardware platform.

Finally, an important motivation for managers in the intended user community was the ability of PD/ICAT to capture the expertise of personnel who were to be transferred to other areas. This factor was key to the ready availability of the experts and in the management's support and dedication of time to this project.

Benefits to the National Aeronautics and Space Administration

Training of astronauts, ground-based flight controllers, and system engineers is a massive task. The best training and the best mechanism for certifying that staff members have met training objectives occur through large-scale, integrated simulations. Unfortunately, these simulations require the support of hundreds of people but deliver training to only one person in each position. The ability of a given trainee to get significant exposure to a particular process is, therefore, quite limited. However, the PD/ICAT system can provide a trainee with virtually unlimited access to training in a specific procedure and ensure that the integrated simulation environment can be used to maximum effect.

Courtesy, NASA

This deployed system demonstrated the capability of intelligent training systems in NASA's operational environment. As a result, a number of similar systems are under development at the space center and other NASA operational centers (Marshall Space Flight Center and Kennedy Space Center). In addition to the impact of this technology on space shuttle training, NASA is supporting its application to future space station training. Because space station training can be a task at least an order of magnitude larger than current space shuttle training, the use of intelligent training systems might be the only way to meet space station training objectives with the available resources. To this end, the Space Station Program is supporting the refinement of the PD/ICAT architecture into a generic training architecture and the creation of a general-purpose development environment to facilitate the rapid adaptation of this architecture to specific training tasks. This latter activity will have a profound impact on the nature of training, not only within NASA but within other government agencies, industry, and the educational arena.

Deployment Process

The conception and initial planning for the PD/ICAT project began in July 1986, and initial knowledge acquisition was complete by December 1986. Code development began in January 1987, and the system was ready for detailed testing and verification by March 1988. During this last period, three complete versions of the system were developed before PD/ICAT was accepted. PD/ICAT has been used for testing and verification by both novice and experienced FDOs since March 1988. The reapplication of PD/ICAT to a Unix-based workstation was accomplished in 1989. The development of PD/ICAT required approximately four person-years and was accomplished by a mixed team from academia, NASA civil service, and private industry. In addition, a number of students contributed to the project during its life. The actual direct cost of the project was approximately $120,000 (for the academic and private-sector portions). The indirect costs of civil service staffing are more difficult to calculate but were approximately $100,000.

Acknowledgments

The authors wish to acknowledge the invaluable contributions of expertise from three FDOs: Captain Wes Jones, USAF; Major Doug Rask, USAF (ret.); and Kerry Soileau. Various students assisted with the knowledge engineering and coding of portions of the user interface and TSM: Tom Blinn, Joe Franz, Bebe Ly, Wayne Parrott, and Chou Pham. Finally, the encouragement and guidance of Chirold Epp (head, Orbit Design Section) and Bob Savely (head, AIS) are gratefully acknowledged. Financial support for this endeavor was been provided by the Mission Planning and Analysis Division, NASA Johnson Space Center; the NASA Office of Space Flight; the NASA American Society for Engineering Education Summer Faculty Fellowships; and a NASA National Research Council Senior Resident Research Associateship. The assistance of Karen Joers in editing the final manuscript is gratefully acknowledged.

References

Brown, J. S.; Burton, R. R.; and de Kleer, J. 1982. Pedagogical, Natural Language, and Knowledge Engineering Techniques in SOPHIE I, II, and III. In *Intelligent Tutoring Systems,* eds. D. Sleeman and J. S. Brown, 227-282. London: Academic Press.

Carbonell, J. R. 1970. AI in CAI: An Artificial Intelligence Approach to CAI. *IEEE Transactions on Man-Machine Systems* 11(4): 190-202.

Hartley, J. R., and Sleeman, D. H. 1973. Towards Intelligent Teaching Systems, *International Journal of Man-Machine Studies* 5: 215-236.

Hollan, H. D.; Hutchins, E. L.; and Weitzman, L. 1984. Steamer: An Interactive Inspectable Simulation-Based Training System. *AI Magazine,* 5(2): 15-27.

Kearsley, G., ed. 1987. *Artificial Intelligence and Instruction.* Reading, Mass.: Addison-Wesley.

Loftin, R. B.; Wang, L; and Baffes, P. 1988. Simulation Scenario Generation for Intelligent Training Systems. In Proceedings of the Third Artificial Intelligence and Simulation Workshop, 69-73. Menlo Park, Calif.: American Association for Artificial Intelligence.

Loftin, R. B.; Wang, L.; Baffes, P.; and Hua, G. 1988. An Intelligent Training System for Space Shuttle Flight Controllers, *Telematics and Informatics* 5: 151-161.

Sleeman, D. and Brown, J. S., eds. 1982. *Intelligent Tutoring Systems.* London: Academic Press.

Sleeman, D. H. 1982. Inferring (mal) Rules from Pupils' Protocols. In Proceedings of the European Conference on Artificial Intelligence, 160-164. Orsay, France.

Wenger, E. 1987. *Artificial Intelligence and Tutoring Systems.* San Mateo, Calif.: Morgan Kaufmann.

Yazdani, M. 1986. Intelligent Tutoring Systems Survey. *Artificial Intelligence Review* 1: 43-52.

Banking and Finance

Analyst: An Advisor for Financial
Analysis of Automobile Dealerships
General Motors Acceptance Corporation

The Intelligent Banking System: Natural Language
Processing for Financial Communications
Consultants for Management Decisions

PFPS: Personal Financial Planning System
Chase Lincoln First Bank N.A. and Arthur D. Little, Inc.

Syntel™: An Architecture for Financial Applications
Syntelligence

TARA: An Intelligent Assistant for Foreign Traders
Manufacturer's Hanover Trust Company

Analyst: An Advisor for Financial Analysis of Automobile Dealerships

Michael A. Hutson

One of the services GMAC offers to approximately 12,000 domestic General Motors (GM) dealerships (and affiliates) is inventory financing. This service is also known as wholesale, or *floor-plan*, financing. In exchange for funds, dealers must adhere to a set of rules, the most important of which is to promptly pay GMAC as vehicles are sold. This adherence is analyzed at least annually.

The process of analyzing a dealership is, in essence, financial risk analysis—not because dealerships are risky businesses but because borrowing and lending money always contains an element of risk. In a typical analysis, a GMAC credit analyst evaluates the risks by examining the dealership's past performance, local economy, and operating ability. The credit analyst then predicts the dealership's likely performance until the next scheduled review. Next, the credit analyst recommends credit lines and suggests ways to reduce risk. Finally, the credit analyst's recommendations are adopted or changed or both by management.

The Problems

To be successful, Analyst must address a number of domain-related problems. First, financial-analysis procedures are time consuming. A single analysis ranges from several hours to a few days depending on the complexity and cooperation of a dealership. On the average, however, each review requires between six and seven hours to complete. Contributing to the length of time credit analysts must spend is the nature of the data: financial statement data are voluminous; prone to error; and in a small percentage of the cases, deliberately misrepresented. This data issue is significant. Its effects are pervasive, from knowledge representation to user interface.

Second, risk-analysis expertise requires years of experience as well as a thorough understanding of accounting principles and financial-analysis theory. Most entry-level employees do not have educations in accounting or finance. Therefore, new analysts tend to concentrate on the mundane tasks of the process, for example, ratio calculations. In addition, although credit analysts quickly become proficient in following established procedures and identifying out-of-guideline situations, they frequently overlook danger signals with respect to the financial condition of a dealership. This problem is exacerbated by a declining experience level in the credit analyst population as a whole. As management retirements increase and business grows (notably in the last five years), numerous promotions from the ranks occur. The ramifications of this declining experience-level problem are that some novice credit analysts tend to be too lenient because they perceive they will lose wholesale accounts to less restrictive GMAC competitors. Conversely, by being overly restrictive, some novice analysts drive high-performing dealerships to competitors.

The third problem is a common one for many businesses: computer-illiterate users. Although nearly all GMAC credit analysts have experience with 327x (dumb) terminals, few have used a personal computer, and prior to this project, none had touched an engineering workstation.

Analyst in Depth

Analyst uses well over 1000 data elements during a review and allows the credit analyst to display several thousand more in forms, tables, and graphs. Most of these data originate from geographically distributed GM and GMAC databases, some of which are not available online.

The central object of the analysis process is the dealer-prepared financial statement. For GM dealerships, this is a four-page form containing a balance sheet; income and expense summary; and detailed information about departmental sales, expenses, and inventories. Most GM dealerships subscribe to a GM service that prepares monthly financial statements from trial balances supplied by the dealerships. Analyst directly accesses this information when so authorized by the dealerships.

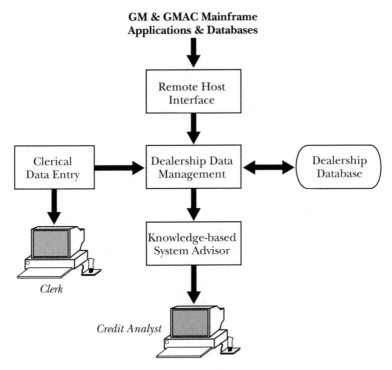

Figure 1. Major Processing Components of the Logical System.

Information about GMAC's experience (rating, credit lines, special programs, monthly retail and wholesale history) with the dealerships is also used during a review. This information is maintained in online databases.

Offline feeds to Analyst from these sources occur nightly or monthly, depending on the source. These data are massaged, organized by dealership, and stored in a central database. Data are kept from the two most recent financial statements so that credit analysts have the option of reviewing earlier information. In addition, information from the previously reviewed and previous year-end financial statements is retained for use during subsequent reviews. Financial data not available from online sources are entered clerically. Actual review processing is done by the knowledge-based system that runs on the Teknowledge, Inc., S.1 tool (figure 1).

Physical Architecture

Analyst was deployed as distributed components. Financial statement information is extracted nightly from an information management system (IMS) database. GMAC experience data are extracted monthly from direct access storage device (DASD) resources. Receipt, processing, storage (in a virtual storage access method (VSAM) file), and preparation for data distribution are performed by batch jobs that run on an IBM mainframe operating under multiple virtual storage (MVS).

The knowledge bases and clerical data-entry components reside in a single Sun Microsystems workstation located at each GMAC branch office. All communications between the workstation and the mainframes are performed over an systems network architecture (SNA) network. To allay concerns about end-user response time and network performance, data are transferred between the host and workstations at night. To complete the process, downloaded data are stored in a local database until the review is completed.

The workstation configuration at the branches is shown in figure 2. The basic branch workstation is a Sun 3/50 with a standard 19-inch monochrome display and a 141 MByte disk. One workstation serial port is connected to a modem-sharing device, which, along with one or more branch control units, accesses the SNA network. The other serial port is connected to an A/B´ (two position) line switch that permits the workstation to alternately access a personal computer and a printer.

The Workstation Platform Software

To help understand the branch workstation software platform, a high-level look is shown in figure 3. Dealership data are stored on an Informix database, with software access provided through Informix ESQL-C facilities. The Informix forms facility presents full-screen displays of the dealership's four-page financial statement. The same form is used to clerically enter dealership financial statement information.

SunLink SNA facilities transfer dealership data (and requests for data) between the host and the workstation and interactively access on-line host applications. The unattended file-transfer operations use the transparent file-transfer (TFT) facility developed by Electronic Data Systems (EDS) Corporation. TFT uses the IBM FTP capability of Sun's TE3278 facility. Credit analysts can access mainframe applications through the M204 tool, which uses TE3278 to provide the 327x terminal emulation.

The User Interface

To make Analyst easy to use, workstation functions are organized as an extendible set of application tools. All the functions run under native SunTools, with separate icons and windows (figure 4) that are displayed after a credit analyst logs in.

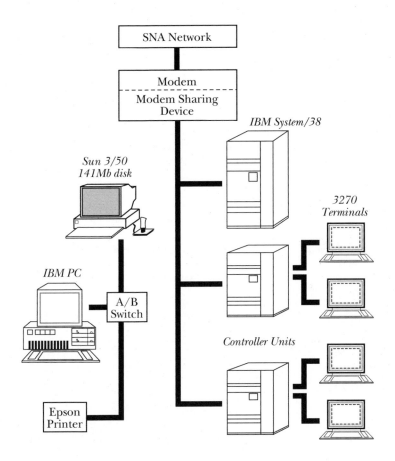

Furthermore, the user interface is the "glue" that joins all of Analyst's major workstation components: knowledge bases, S.1, Informix database, and 327x terminal emulation. Conceptually, the user interface is organized into layers. The top layer is a program (actually, just a UNIX shell script) that starts up—in the background—each of the tools. The second layer comprises the tools themselves, each consisting of a base frame and screen(s). Each screen is composed of a set of building blocks that make up the bottom layer.

Knowledge Representation

Expert dealership risk analysis combines data abstraction and evaluation, association of problems with corrective actions, and refinement of

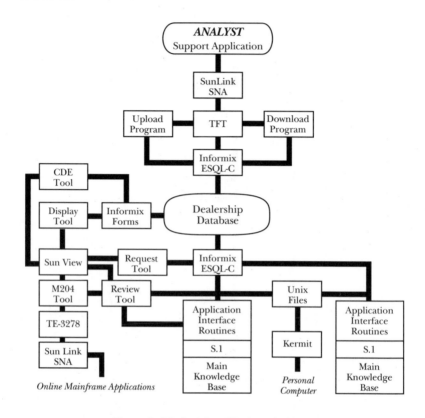

Figure 3. Workstation Platform Software.

recommendations (figure 5). In the initial step, each significant piece of financial statement data (or calculation involving several financial statement elements) is abstracted into more significant data features. Implemented in the form of rules and constraints, this step allows the expert to reason with symbolic, rather than numeric, data.

The initial step also involves an evaluation of the numeric data; the symbolic abstractions generally indicate when the financial statement data (or calculation result) are out of the guidelines. For example, the values concerning used car inventory level and rate of sales become statements about the dealership's inventory status. Thus, a 73-day supply of used cars can be abstracted as an "overstock of used cars" (figure 6).

Once the data are in symbolic form, the credit analyst interprets the data characteristics and further abstracts out-of-guideline situations into risk situations. Here, reasoning is accomplished with more complex rules (because more judgment is involved) than in the initial ab-

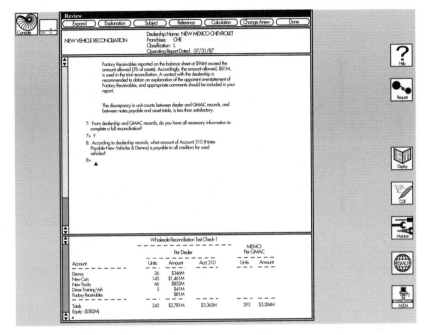

Figure 4. Analyst Icons and Review Tool Window.

straction step. To continue the example, further analysis might conclude that the dealership (and GMAC) is at risk because of the excessive investment in used car inventory.

In the next step, risk situations that have been identified are associated with potential corrective strategies. For example, if a dealership has an overstocked used car inventory, one recommendation is to liquidate the inventory.

In the final step, the recommendation is refined using additional analysis and, possibly, requiring the analyst to gather more information. To complete the example, selling just the used cars on hand for over 30 days to wholesalers might remedy the risk situation.

The Knowledge Bases

In the original design of Analyst, only one knowledge base existed. As the system matured, it became evident that credit-analysis knowledge could be used in three different ways: (1) to verify that the clerically entered data met a minimum set of requirements, (2) to perform a preliminary analysis on the data received from the host before a consultation commenced, and (3) to conduct an interactive consultation

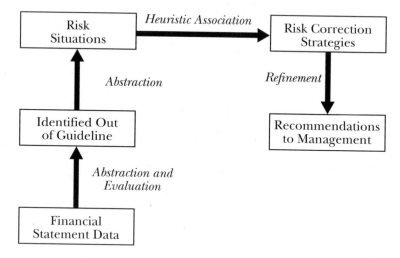

Figure 5. Knowledge Representation Steps.

with the branch credit analyst. Accordingly, three knowledge bases were developed. In this article, only two of the knowledge bases are discussed: the front end (noninteractive) and the main (interactive).

As mentioned earlier, data reliability is a concern of credit analysts. Because data are the starting point in knowledge representation (figure 5), the front end performs the important function of advising the credit analyst about missing, inconsistent, and questionable data.

The front-end knowledge base was designed to remove data validation and routine calculations from the main knowledge base. The advantages of this design are a smaller, more maintainable main knowledge base and faster running consultation. Again, rules and constraints were used to model the expert's reasoning.

Among the activities of the front end are the following:

- To perform calculations and check the sensibility of the results
- To perform screening tests to check for the validity of balance sheet items
- To check if these two tests yield significant problems
- To compare the current and previous financial statements for significant changes
- To summarize and print the findings
- To set an intensity level that determines the depth of questioning during the consultation

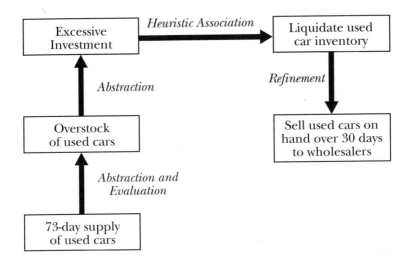

Figure 6. Knowledge Representation Example.

As an example of a front-end activity, setting an intensity level is achieved by determining the attribute high.anxiety?[r], which indicates that the dealership might be in trouble. This boolean attribute is set only when one of several rules is fired. The control block det.high.anxiety? controls the processing, including storage of the results in the local database. The control block attribute and a typical rule are shown in figure 7.

The main knowledge base runs the consultation and produces the review reports. It contains over 3500 objects organized into layers. The organization of this knowledge base was partly motivated by the observation that a large amount of the area-specific analysis (for example, cash, new vehicle inventory) is nearly independent of other areas. Further, the same kinds of processing were done for each area. Although S.1 does not provide an oops capability, a general object-oriented approach was chosen for the high-level control design to achieve simplification in development, maintenance, future extension, and—to appease end users—enhanced explanation facilities.

The highest-level consultation control blocks are primarily concerned with the sequence of processing as perceived by the user. The sequence of events in the topmost control block is as follows:

1. Create the class instance for the review data.

2. Get the name of the dealership to review (passed in from the user interface.)

```
DEFINE CONTROL.BLOCK det.high.anxiety?
/* fe.general.ctl */
::ARGUMENTS a:attribute, r:review
::INVOCATION determination
::BODY          begin
                vars dummy.b:boolean
                seek a[r] by rules;
                if high.anxiety?[r] known then
                    dummy.b:=db.put.b(high_anxiety,
                    high.anxiety?[r])
                else dummy.b:=db.put.null (high_anxiety);
                end
END.DEFINE

DEFINE ATTRIBUTE high.anxiety?
/* fe.general.ctl */
::DEFINED.ON r:review
::TYPE boolean
::MULTIVALUED false
::LEGAL.MEANS {try.rules}
:: DETERMINATION.BLOCK det.high.anxiety?
END.DEFINE

DEFINE RULE rul.intensity.net.loss
/* fe.general.rul */
::APPLIED.TO      r:review
::PREMISE         (db.get.i(net_profit_aft_tax) < 0
::CONCLUSION high.anxiety?[r]
::JUSTIFICATION "if there is a net loss after bonuses and taxes, this review
    should be conducted at the maximum level of intensity."
END.DEFINE
```

Figure 7. A Control Block Attribute and a Typical Rule.

3. Fetch the associated dealership data row, and determine whether the user wants to review this dealer (during which the results of the front end are displayed).

4. If the user accepts the validity of the data and elects to proceed, invoke the initiate.review control block, which controls the following processing:

• Determines from the user which type of review to perform

• Performs specialized processing as dictated by the review type

• Determines whether the review should be conducted at a high or low level of detail (intensity)

• Saves the reason for review in the database for later uploading to the host

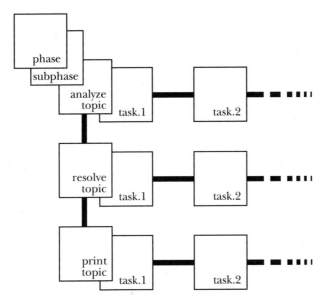

Figure 8. High-Level Main Knowledge Base Structure.

- Invokes the review type–specific control block that directs the sequence of consultation phases to be performed

5. Updates the database.

Subsequently, the conduct of the consultation is determined by one of the eight review types. In turn, they define which area-specific analyses are to be performed. For example, the beginning section of the regularly.scheduled.review control block looks like figure 9.

Knowledge base layers make up the rest of the framework, beginning with phases (figure 8). *Phases* reflect the general nature of the processing currently being done (for example, profit-and-loss analysis). The current phase is always displayed on the screen to provide a frame of reference for the credit analyst. Phase-control blocks conditionally invoke subphases that are relevant to the type of review (for example, regularly scheduled, dealer loan request) to be performed. Subphase control blocks, in turn, invoke the area-specific analysis control blocks, or topics. Three forms of topic-level control blocks execute approximately 10 different functions, including performing analysis ("analyze topic"), determining recommendations ("resolve topic"), and generating reports ("print topic"). Invoked by topics, task-level control blocks perform a specific set of activities, some of which require user interac-

```
DEFINE CONTROL.BLOCK regularly.scheduled.review
/* top.level.ctl */
/* This control block is invoked from initiate.review in the case of a regu-
    larly scheduled formal review. */
::ARGUMENTS        r:review
::TRANSLATION      "control regularly scheduled review
    sequencing:
::INVOCATION internal
::BODY             begin
                   invoke analyze.character.phase (r);
                   invoke reconcile.ws.inventory.phase (r);
                   invoke analyze.profit.and.loss.phase (r);
                   (and so on...)
```

Figure 9.

tion. The scope of each task is limited so that explanations available to the end user make sense. Because explanations are a critical and unique aspect of Analyst's delivered capabilities, they are discussed in greater detail in the next subsection.

Explanation Facilities

The challenge in providing useful explanations was the credit analysts themselves; some would be experienced and others complete novices. To meet the needs of both (and those in between), multiple types of layered explanations were created in two broad categories: context sensitive and context independent.

All explanations are canned text located in a subdirectory on the workstation. These text files were created and are maintained solely by GMAC, which considered the native S.1 explanation facilities insufficient for end users. In the following examples, compare S.1 (top) and Analyst text when executing a "why" command:

"I asked about the date of the last wholesale inventory in order to apply an.wholesale.inventory.since.review?.1. An.wholesale.inventory.since.review?.1 is used to determine whether there has been a wholesale inventory completed since the last formal review.

"The most fundamental asset on a dealership's lot is the inventory the business sells. Some dealers own this inventory outright, but most of them finance their new vehicles, and GMAC provides that financing. Unfortunately, the process of reconciling the new vehicle inventory asset against the corresponding liability is complicated by the many different accounts the dealer can use in the day to day flow of business.

Type	Functionality (Answers the question:)
Help	What do I do next? What are the implications of my answer?
Theory	What accounting theory (or principles are at play?)
Syntax	What is the set of legal answers to the current question?

Table 1. Context-Sensitive Explanations Tied to Attributes.

But I can help you sort through this complicated process, provided you obtain all the necessary information for the wholesale reconciliation."

Explanations are available only when a task-level control block contains interactive code. In other words, the consultation must be at a question for a credit analyst to access any explanation type. From the user's standpoint, all explanations are obtained by simply clicking a button or using a pull-down menu. Internally, the user interface pipes a command string to an S.1 external function that retrieves the text and places it in the correct window. The mechanisms for retrieving the canned text differ by the type of explanation sought.

Context-sensitive explanations are tied either to an attribute (table 1) or to the structure of the knowledge base (table 2).

Text file location and layering are achieved by simply adding a unique file-name extension ('.hlp1', '.hlp2', '.hlp3') to an attribute name. As the user asks for additional help, the S.1 external function routine increments a counter and retrieves the appropriate file from the explanations directory. Three levels of help are available for every question, each level containing more detailed explanations. Only one layer is implemented for theory and syntax.

The mechanism for locating text differs for the next two explanation types. Among the arguments passed to a task-level control block are why and situation keys. These keys provide the indexes for looking up the canned text. A stack is kept for each of these two explanation types. Thus, as lower-level tasks are invoked, keys are stacked. When a credit analyst repeats the need for one of these explanation types, a key is removed from the stack. Therefore, each successive request returns an

Type	Functionality (Answers the question:)
Why	What am I doing here? What is going on at this point?
Situation	How did I get here?

Table 2. Context-Sensitive Explanations Tied to Knowledge-Base Structure.

explanation more general than the previous one—the knowledge base structure is climbed from bottom to top.

Context-independent explanations are shown in table 3. For these explanations, the text-retrieval method differs slightly because knowledge base attributes are not involved. Specifically, the credit analyst determines (by selecting an item from a pull-down menu or highlighting text in the consultation) the "name" used as an argument to the command string, which is passed to the S.1 external function.

Project Management

For a dealer review application to be considered successful, GMAC expected to accrue a number of tangible and intangible benefits. The minimum payoff from tangible benefits was expected to be $2 million per year. Depending on the ability of Analyst to reduce dealer losses resulting from default, the utilization of the system, and the state of the economy, the system should increase the payoff:

- Tangible Benefits
  - 50 percent reduction in time to perform reviews
  - Reduced effort for internal compliance audits
  - Elimination of planned costly training course
  - Reduction in losses resulting from dealer default
- Intangible Benefits
  - Standardization of dealer review process
  - Thorough, consistent analysis
  - Enhanced performance of credit analysts
  - Additional time for the difficult reviews
  - Value-added service to dealers
  - Method for training credit analysts
  - Ability to incorporate new analysis techniques

Type	Functionality
Subject	Explains key business concepts in the form of definitions of items such as materiality, intangible assets
Reference	Provides online access to GMAC operations and credit analysis manuals
Calculations	Displays the general formula, instantiated formula for the dealer under review and the result of the calculation

Table 3. Context-Independent Explanations.

- A precursor to
  - Analyzing any automobile dealership
  - Analyzing other types of businesses
  - Analyzing megadealerships
  - Constructing models of business types
  - Marketing certain analysis modules
  - Incorporating new analysis techniques

Project Life Cycle

Late in 1985, GMAC focused on an AI solution to automating the dealer review process. The study expanded when GMAC approached Teknowledge, Inc., of Palo Alto, California. Following conceptual and engineering validations, prototype development began.

At the completion of the prototype in December 1986, GMAC conducted numerous demonstrations of Analyst for its executives, managers, and credit analysts. Because of the positive feedback from the demonstrations and written system evaluations, knowledge base development resumed in January 1987.

System engineering work was undertaken by GMAC, Teknowledge, EDS, and American Management Systems of Alexandria, Virginia, in March 1987. The following January, the system was installed in a single pilot branch. Eight branches were added to the pilot program in June 1988. National rollout commenced in July at the rate of 16 branches per week. By December, all 230 systems were installed and operational.

Results

At the time of this writing, credit analysts had already performed 12,000 dealer reviews using Analyst. Feedback from them over the months has been particularly rewarding. Although generally inexperienced with windows and mice, credit analysts quickly overcame their initial apprehension. They think the system is easy to learn and use. Beyond their familiarization with the personality of Analyst, the normal learning curve for a good understanding of system concepts seems to be about three weeks. Users are then comfortable enough with the system to trust it—and ask questions about its logic.

Objective feedback was obtained too. In a recent branch survey, users verified that many of the intangible benefits were being met—some exceeded. Although the full tangible impact of the system cannot be realized for several months, most users indicated that the system was already providing measurable time savings.

Acknowledgments

I am particularly grateful to a unique group of hard-working people—Analyst survivors—without whom this project would still be in development and would have been less fun. You know who you are. This project is a credit to GMAC executives, who were willing to take a fair amount of risk.

The Intelligent Banking System: Natural Language Processing for Financial Communications

Kenan Sahin and Keith Sawyer

International banking relies heavily on the electronic transfer of messages for basic transactions. Until the mid-1970s, messages were transmitted over the telex carrier networks, as natural language text. In the mid- to late-1970s, the major international banks developed several industry-wide structured formats to represent the most common banking messages, such as funds transfers. These formats allowed the banks to develop computer software that could automatically process the structured transaction, precluding the need for manual intervention. (This development parallels the more recent moves to electronic data interchange [EDI] in other industries.)

Despite the widespread success of this strategy in reducing processing costs and increasing bank productivity, a significant minority of the international message traffic remained natural language text. This traffic still required costly and error-prone manual processing. Proficient operators needed a significant understanding of international banking transactions, creating high training costs and limiting staffing flexibility.

Because of the need to process English-text input and incorporate a significant amount of domain expertise, traditional programming techniques were inadequate. AI technology was identified as the appropriate solution. AI offers two groups of techniques that Intelligent Banking System (IBS) uses: natural language processing techniques, and rule-based expert system techniques.

The goal of IBS is to use a combination of these techniques to allow the computer to scan and understand a natural language text message. Automating the task in this manner would reduce banking costs, increase operator productivity, and reduce the chance of manual error. The task seemed appropriate for this technology, because the application satisfied many of the accepted criteria (Davis 1982 and Prerau 1985):

- The domain is characterized by the use of expert knowledge, judgment, and experience.
- Conventional programming solutions are inadequate.
- Recognized experts solve the problem today.
- The completed system is expected to have a significant payoff for the corporation.
- The task requires the use of heuristics, or rules of thumb.
- The task is neither too easy nor too difficult.
- The system can be phased into use gracefully.

IBS System Description

IBS was developed on a custom basis by Consultants for Management Decisions, Cambridge, Massachusetts, (CMD) for Citibank, New York. IBS was originally developed for the funds transfer class of banking messages. The methods and approaches applied to this domain proved readily extensible to other types of banking messages, and IBS has since been extended to several other message domains. Thus, IBS is actually a full family of applications, including the following modules:

- Funds transfer message processing
- Letter of credit issuance message processing
- Letter of credit reimbursement message processing
- Funds transfer problem inquiry message processing
- Message classification (involving an analysis of all telex traffic to determine which domain-specific module is appropriate)
- Test-key parameter identification

All these modules are fully integrated in various production environments.

Several of the modules are implemented onsite at Citibank. These systems are fully embedded in the existing bank processing systems and are processing live telex traffic on a daily basis. Several of the modules are also installed in the data processing facilities of two telex carriers: Western Union-ITT, and TRT. These telex carriers are offering IBS message enhancement as a service to their banking customers. IBS is fully integrated with the production processing environment at these sites as well.

Because it was the first to be developed, the funds transfer application is perhaps the most mature system. This system is accompanied by the Intelligent Banking Workstation (IBW), which allows an operator to review an IBS-processed message using a window-and-mouse-based interface. IBW allows intelligent entry, which provides the user with full access to the knowledge capabilities of the system. IBW also provides the user with explanations of the various actions taken by the system during the parsing and resolution phases. IBW further increases operator productivity by allowing the operator to make use of partial information identified by the system.

The other applications are not currently accompanied by their own user interfaces but instead are integrated so closely with the existing production systems that the standard terminal interfaces of the bank processing systems can be used.

The funds transfer module has been in production since 1985. This effort followed the original prototype, which was completed by CMD in mid-1984. The total calendar time for the effort to transform the prototype to full production was approximately eight months. This initial production system was implemented on a dedicated Lisp machine and networked using a custom-developed protocol. This implementation proved to be inappropriate for full production, and the system was ported to the VAX environment in 1986.

The later applications began during 1986 and 1987 and were each designed from the beginning for production implementation on the VAX platform (only a brief prototype phase was included).

Technical Description

The knowledge domain for IBS is the reading and translating of English-text messages into a structured format. The messages are sent to Citibank electronically by way of internal, proprietary networks, and external telex networks. Depending on the type, the message can be from 80 words to several pages long.

Message Characteristics

An example from the funds transfer area demonstrates the unique nature of these messages. A typical money transfer telex (with the bank

```
ABC3456 FGH0055
NYAAA
.CAREXEX 1415144
TEST 12345 AMT 40000
FOOBAN 45345GF
FROM: FOREIGN BANK, PARIS
TO  : BIG BANK NEW YORK, U.S.A.
TEST: 12345 PLS DEBIT OUR ACCT NO 1234567 WITH YOURSELVES
AND PAY VALUE 21.03.89 THE SUM OF USDLRS 40.000,-
TO FOREIGN BANK NEW YORK, 1 MAIN ST. N.Y. 10000
FOR OUR ACCT NO 4567
   WITH THEM.
THANKS/REGARDS
OUR REF: TT MM 3333    STOP.
MSG NO 333 FOREIGN BANK,
   PARIS
FOOBAN 45345GF
```

Figure 1. A Typical Money Transfer Telex.

names fictionalized) appears in figure 1. The subset of English used in these messages is highly terse and abbreviated. The people typing in the messages are under time pressures, so abbreviations and typographic mistakes are common. Many of the messages are entered by people for whom English is a second language. Often, information that is necessary for the recipient but not required of the sender, is omitted to reduce the sender's message entry time. For example, the name of a bank is often specified without the corresponding account number. In many messages, information that is not needed by the recipient is supplied ; this information must be ignored.

The domain is such that a direct mapping from individual phrases to structured values is not possible. The structured values depend on the context of the entire message. Some structured values depend on several different phrases in combination. Some values can depend on a particular combination of yet other structured values. Thus, the possible combinations of situations resulting in a given value are large. Application domains in which combinatorial effects become significant usually do not submit to a cost-effective, traditional programming solution. These complexities are an indication that AI techniques might be appropriate.

In addition to these domain requirements, the production environment required that each message be processed in under 60 seconds.

```
:CLASS:
:TYPE:FUNDS
:TEST:
:WRD:12345 :AMT:40000 :FROM:FOREIGN BANK
:MAP:
0000 00BSTELEXXAXXX00000
0000 44IBSTELEXAXXX00011
:202 02
:20:TT MM 3333
:32A:890321USD40000,
:53D:/1234567
FOREIGN BANK
PARIS,FRANCE
:57D:FOREIGN BANK
NEW YORK, NEW YORK
:58D:/4567
FOREIGN BANK
PARIS,FRANCE
—
```

Figure 2. The Output of the IBS System.

The output of the IBS system, for the telex shown in figure 1, is shown in figure 2. This output conforms to the rigid format specifications required by automated transaction processing systems. The format used in this example is known as SWIFT.

System Design
IBS makes use of a hybrid approach, borrowing ideas from several significant concepts in computational linguistics and rule-based expert systems. The abbreviated version of English found in these messages led to the use of a flexible parser approach (Hayes and Mouradian 1981). The system combines elements of case-frame grammars (Fillmore 1968) and semantic grammars (Hendrix 1977) to arrive at the final linguistic formalism.

This formalism was designed using a variation of the augmented transition network (Woods 1970) to build semantic units. Each semantic unit is responsible for the identification of one key piece of information from the telex. As information is identified, it is stored within the semantic unit.

The characteristics of our formalism satisfied the domain requirements:
1. The formalism was capable of identifying single phrases and incomplete sentence fragments.

2. The formalism was able to identify useful information and ignore irrelevant information.

3. The formalism provided for the identification of abbreviations and misspellings.

4. By taking maximum advantage of the domain constraints, the formalism provides for highly efficient processing of the English text.

In addition to this linguistic formalism, we employed a rule-based expert system to incorporate domain knowledge. The expert system receives input from the semantic unit values. It is used to make decisions based on overall message content, to infer values using combinations of phrases, and implement constraints among different structured values. This expert system was also custom developed to achieve production-level speed. In the current version of IBS, the rules were rewritten directly in Lisp code, resulting in a tenfold performance increase.

The true originality of IBS lies in its unique blend of several different research concepts to result in a system that satisfies a specific business goal. Despite the use of these fairly advanced concepts, IBS can still process an average telex in 30 seconds on a machine as small as an IBM personal computer.

The Intelligent User Interface

IBS was designed to process a message fully, then pass the message and the corresponding structured information to a user-edit interface. IBS identifies an average of over 80 percent of the structured information. An operator must complete the remaining structured information, usually one or two values.

The data-entry stations at many banks cannot support the display of both the message and the structured equivalent. These interfaces were designed to be used with a printed copy of the message and provided only for structured value entry. Designing IBS to print the telex for these operators would have considerably reduced the cost-effectiveness of the process. Instead, we implemented an intelligent assistant, the Intelligent Banking Workstation (IBW) as a companion to IBS. IBW was conceptualized as a low-level assistant to a human operator that would provide much of the processing expertise, freeing the operator to perform higher-level conceptual activities (Rich and Waters 1981).

IBW employs mouse cursor control, multiple windows, and pop-up menus and windows to improve operator productivity. Two primary windows are displayed: one containing the original message and one containing the structured values identified automatically. The mouse can be used to mark a region of text in the message window and move this text into one of the structured values.

Incomplete or ambiguous values identified by IBS are made available to the user through pop-up windows. One such window is for English-text notifications of problems encountered during processing. A second window contains suggested values that are each mouse selectable. For example, if several branches are found in the IBS bank database for the name Credit Suisse, the notification window would say "Several branches found for Credit Suisse," and the suggested values menu would display each of the branches, with the corresponding city and account numbers. This mechanism allows the user to benefit even when IBS cannot uniquely identify a value.

The linguistic and domain knowledge used in the automatic processing is also available to the user. For example, when a region of text is moved to a structured value using the mouse, the user can request intelligent processing for this text. The parsers and domain rules are then invoked to process the text. A correctly processed value is entered by IBW. In addition, other values that might have been affected by this change are flagged with a notification for the user.

This intelligent user interface is a significant value-added component of IBS. The power of the intelligent assistant concept—employing mouse cursor control, multiple windows, and pop-up menus—significantly increases the productivity of the users. Providing a broad interface between the users and the intelligence in the system results in maximum value for the knowledge engineering effort.

Success Criteria and Payoff

The application was determined to be successful if over 80 percent of the information in an average telex was identified automatically and if each telex could be processed in a short-enough period of time to be cost effective (approximately 30 seconds for most message types). Each of the many IBS modules has met or exceeded this criteria. At this success level, implementation of the module is considered cost effective.

The payoff for each of the modules varies depending on the specifics of the installation site. Because each module is installed in varying configurations (telex carrier site versus money-center bank), the cost savings vary. Generally, the savings can be characterized by reduced head count, increased customer satisfaction, and lower cost resulting from data-entry error.

Acknowledgments

The authors would like to emphasize that IBS is the result of a collaborative effort involving persons too numerous to mention. The authors would like to recognize the key efforts of John Hodgkinson at CMD and the critical efforts of all involved at Citibank.

References

Davis, R. 1982. Expert Systems: Where Are We? and Where Do We Go From Here?, Massachusetts Institute of Technology. AI Memo No. 665.

Fillmore, C. 1968. The Case for Case. In *Universals in Linguistic Theory,* eds. E. Bach and R. Harms. New York: Holt, Rinehart, and Winston.

Hayes, J. H. and Mouradian, G. V. 1981. Flexible Parsing. *American Journal of Computational Linguistics* 7 (4): 232-242.

Hendrix, G. G. 1977. The LIFER Manual: A Guide to Building Practical Natural Language Interfaces, Technical Note 138, SRI International.

Prerau, D. S. 1985. Selection of an Appropriate Domain for an Expert System. *AI Magazine* 6 (2): 26-30.

Rich, C., and Waters, R. C. 1981. Abstraction, Inspection, and Debugging in Programming, AI Memo No. 634, Massachusetts Institute of Technology.

Woods, W. A. 1970. Transition Network Grammars for Natural Language Analysis. *Communications of the ACM* 13: 591-606.

PFPS:
Personal Financial
Planning System

*Kyle W. Kindle, Ross S. Cann,
Michael R. Craig, and Thomas J. Martin.*

"How can I achieve, as closely as possible, my lifetime financial goals given my limited resources?" The Chase Lincoln First Bank Personal Financial Planning System (PFPS) was designed to solve this problem, defined from the client's perspective. PFPS provides objective, affordable expert financial advice to individuals with household incomes ranging from $25,000 to $150,000 and up. This system was developed by Chase Lincoln First Bank to supplement the use of high-cost bank personnel in the strategically important but heretofore unprofitable area of personal financial planning services.

The system is of great value to the bank. Relationships developed with customers through the planning process often lead to additional bank business related to the implementation of the planning recommendations. The incremental cost of producing a plan using PFPS is more than recovered as a client fee. Thus, what has generally been for banks a loss leader for up-scale clients has become, using PFPS, a cost-effective opportunity to serve a much wider client base.

The planning system eliminates the individual biases of the human planner and far exceeds their ability to consider the wide range of information that might—and should—affect a financial decision. PFPS integrates planning modules with expertise in the following areas: investments, debts, retirement savings and settlement of retirement plans, education and other children's goal funding, life insurance, disability insurance, budget recommendations, income tax planning, and savings to achieve miscellaneous major financial goals

PFPS develops a set of strategies that enables the client to attain the established goals or make reasonable concessions that trade off among conflicting goals based on client priorities and the timing and amounts of any shortfalls. For example, the age of retirement might conflict with the postretirement standard of living. One goal, or a combination of both goals can be modified depending on the amount of the shortfall and the client-provided priorities.

Strategies can include the use of alternative savings and investment options, such as custodial accounts; company-sponsored retirement plans; individual retirement accounts (IRAs); tax-exempt investments; and, where necessary, debt. The system maintains a balanced cash flow while it optimizes investment returns and the tax consequences of selected strategies.

The client provides data regarding sources of income, level of expenses, assets, liabilities, insurance coverage, employee benefits (retirement and insurance), risk tolerance, and lifetime goals ranked according to priority through an extensive questionnaire and an interview with a financial planner. Goals can include the standard of living now and during retirement, retirement age, adequate levels of insurance protection, college or other goal funding for the children (wedding, home down payment, and so on), and provision for miscellaneous major goals. The system determines what level of these goals can be achieved and develops a customized strategy for their achievement.

The system is driven by the client-supplied data together with a set of parameters that describe the external environment, for example, types of investments available, types of loans available, rates for general inflation, education and real estate inflation, and insurance costs. This planning parameter file contains on the order of 15,000 pieces of data.

Candidate solutions are generated and examined by PFPS to ensure that all aspects of the client's cash flow and asset levels are in balance over the entire planning period. The planning period often covers 50 years and is selected to ensure that less than 5 percent of the clients outlive the plan.

The output from the system is a final report for the client, ranging in size from 75 to 100 or more pages with text, tables, and graphs that clearly explain the recommendations and the advantages of these strategies. The final report is divided into three sections: (1) specific

FINANCIAL STRATEGIES FOR JOHN & MARY RIGHT

CHASE LINCOLN FIRST BANK, N.A.   November/1989

Your Goals

Our projections of the achievable level of each of your goals, based on following all of the recommendations in this plan, are outlined in the following table.

Your Goals for Personal Financial Planning
(1989 Dollars)

Priority Rank	Goal	Desired Objective	Projected Achievement Based on Plan	Percent Achievement Based on Plan
1.	Retirement Age/Year			
	John	57/2002	57/2002	100%
	Mary	56/2002	56/2002	100%
2.	Funds for Retirement Living Expenses (annual after-tax)	$30,565	$30,565	100%
3.	Maintain Standard of Living – Current Budget	$24,440	$23,218	95%
4.	Annual Income if Disabled*			
	John	$30,530	$30,530	100%
5.	Funds for Children	$82,000	$69,700	85%
6.	Annual Income for Survivors			
	Mary	$19,624	$17,662	90%
	John	$19,624	$19,624	100%
7.	Boat (1990)	$25,000	$20,000	80%

*For Other Protection, see "Other Insurance" section of Plan.

Figure 1. Sample Report Page Showing Goal Achievement

action items for the next year or two, (2) recommendations for the next three to five years, and (3) an appendix containing recommendations and supporting charts for the balance of the client's life. Figures 1 and 2 show two pages of output from a typical client plan.

Generic recommendations that leave the client wondering what to do are avoided. Instead, yearly recommendations are provided that are specific to the client situation. For example, rather than describing the general benefits of participating in defined contribution plans, the plan might recommend that the client contribute 6 percent to a 401(k) plan starting in May 1990 and increase it to 10 percent in Jan-

FINANCIAL STRATEGIES FOR JOHN & MARY RIGHT

| CHASE LINCOLN FIRST BANK, N.A. | November/1989 |

Recommended Risk and Return Projections

The next graph shows the expected performance of the investments we recommend. We have factored in a risk level that is consistent with your stated tolerance for risk, and the performance reflects a rate of return consistent with our recommendations. The table on page A.51 of the Appendix has the actual numbers which support these projections.

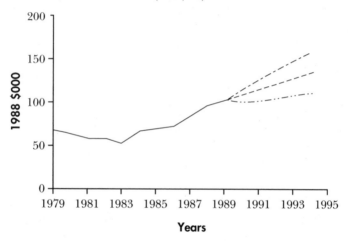

Recommended Portfolio Risk and Return Projections
Investment Assets After-Tax
(1989 $000)

Figure 9. Recommended Portfolio Risk and Return Projections: Historical Value (solid line), High Range (dash-dotted line), Low Range (dashed and dots line), Expected Value (dashed line).

Figure 2. Sample Report Page Showing Risk-Return Envelope for Recommended Client Portfolio

uary 1991. All recommendations are fully supported by charts showing detailed sources and uses of cash for the first five years of the plan and summary charts covering the client's entire life.

Development History

PFPS and the personal financial planning service have been under development for the past five years at a cost to Chase Lincoln First of close to $5 million for system development, product development, and marketing promotion. The architecture and initial version of the system were developed by Thomas Martin, Kyle Kindle, and others at the Arthur D. Little AI Center between 1984 and 1986 under Ross Cann's supervision at Chase Lincoln First.

The initial version of the system, written in Prolog, used a simple form of generate and test to construct, evaluate, and dismiss trial solutions. PFPS was then redesigned and written in ZetaLisp to enable the creation of a full-scale, real-world solution.

Between 1986 and initial market testing in late 1987, Cann, Kindle and Craig completed the implementation of the planning modules that deal with the individual areas of financial planning expertise. Additional work was also undertaken at Chase Lincoln First to enhance the goal-modification process and permit larger areas of the search space to be pruned to achieve substantial run-time improvements for client plans requiring goal reductions.

Controlled market and product testing continued through early 1988, and personal financial planning was rolled out on a limited basis in the Rochester, New York, area in September 1988. Preliminary market studies indicate a high level of satisfaction among those people who have received financial plans.

System Overview

The PFPS system components are implemented as an embedded system on an IBM 4300 series mainframe under VM/CMS and on a Symbolics 3600 series Lisp processor (figure 3). The overall system is controlled by a group of virtual machines on the IBM 43xx. Separate virtual machines exist for data entry and verification (KEYENTER), composition of the final report (COMPOSER), production supervision (PRODSUPR), control of the customer database (DBCNTL), and communication with the Symbolics 36xx (PLANNER). Additional virtual machines exist for text maintenance (MAINT) and control of the planning parameter file (PPFMNT).

The separation of duties among virtual machines on the IBM side reflects the extension of the object-oriented programming paradigm from the Symbolics realm into the IBM. In addition, a high-volume production environment is made possible by this design because any number of data-entry virtual machines can be added to the system at an extremely low cost (at local or remote locations).

Data supplied by the client are key entered and verified on KEYENTER and then put through a series of several thousand edit checks to ensure the internal consistency of the data. Because the client data vary from individual to individual, anywhere from 100 to 2000 data ele-

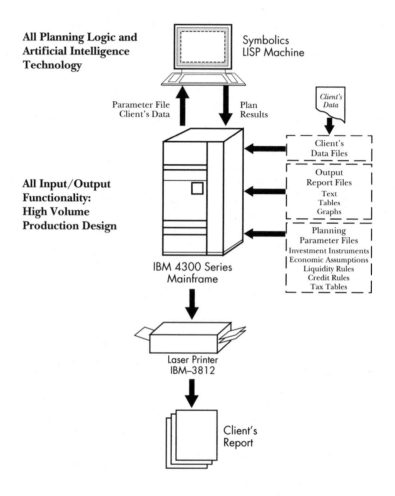

Figure 3. PFPS System Architecture Overview

ments are supplied. After the client data have been approved for processing and released by PRODSUPR, they are downloaded to the Symbolics by the PLANNER virtual machine.

The Symbolics system is connected to the IBM mainframe in a master-slave relationship using an application-level protocol defined on top of RS232C. The Symbolics operates as a slave to the PLANNER virtual machine, which initiates batch processing on request. Every 30 seconds, PLANNER interrogates the Symbolics about its state and transmits client data, waits for planning to be completed on the Symbolics, or receives plan output files. The Symbolics solves the plan as a single, low-priority process with a high-priority process monitoring the serial communications line. The IBM system can enforce a system time-out, which is designed to guard against runaway plans. If the Symbolics crashes in the course of a plan, the IBM reboots the Symbolics and proceeds to the next plan.

This innovative, direct, real-time connection antedates such well-known applications as the authorization assistant of American Express. The design allows the planning code to be maintained and enhanced using high-level Lisp, macros, and other features on the Symbolics. Rewriting the system in an IBM-supported language would have imposed far greater development and maintenance costs. From an ongoing maintenance and development standpoint, PFPS must react quickly to changes in prevailing tax laws, investment types, and so on. Further, this design facilitates the scaling up or down of the processing power of PFPS through inexpensive upgrades to, or the addition of, Symbolics workstations.

All the inferencing and planning is performed on the Symbolics and is written entirely in Lisp, making extensive use of Flavors. PFPS has also been successfully ported to and run on a Texas Instruments (TI) Lisp machine as part of a compatibility and performance study, although no TI machines are included in the current configuration. The planning logic, which is described in greater detail later, includes more than 200,000 lines of Lisp code.

On completion of Symbolics processing, a results file is uploaded to the IBM, merged with text files, and composed using IBM's DCF/Script product for final printing of the client report on an IBM 3812 laser printer.

Object-Oriented Database

After downloading of data from the PLANNER virtual machine to the Symbolics file system, the data are loaded into an object-oriented database that serves as the blackboard for the PFPS planning environment. The database structure, implemented using Symbolic's Flavors, follows the theories of semantic data modeling described in Curtice and Jones (1982).

The requirements of PFPS were such that the database of a traditional blackboard model was extended. As outlined in Weinreb et al. (1987), an object-oriented database extends the features of an object-oriented language by offering persistence, transactions, and sharing between workstations. Moon (1986) describes the approach in the Symbolics Lisp operating system (Genera) to use the Flavors extensions to Lisp for object-oriented programming.

Flavors alone, however, did not offer features generally required in the information system industry, particularly the financial community. For example, Flavor's error checking and enforcement were weak. The authors required a method for enforcing strong typing on objects so that project collaborators would have clear conventions by which to program. The subsequent design process produced an exhaustive record structure consisting of a determinate schema mechanism that could embody a system data dictionary. The schema allowed many programmers and users to see the system's data elements, which consisted of such items as the client's name, the account, the account type, yearly income statements, and the client's financial goals (figure 4).

The memory requirements of the database that is created on the Symbolics is large and unpredictable in advance of a client run. Some of the complex schema, such as an individual's income statement, have many instantiations, one for each person, in each year that a plan is run. An early version of PFPS put the entire database in a temporary area that was then flushed clean between clients. However, the difficulty of keeping operating-system objects out of the temporary area made this solution unworkable.

As additional logic was added to the system, the memory requirements increased to the point where running a single plan could use as many as 100 MBytes of memory, including garbage. As a result, current memory management makes use of the most extreme form of garbage collection: automatic booting of the Symbolics between client plans by command from the PLANNER virtual machine. Booting between plans takes less than five minutes and is more than offset by the reduction in planning time attained by the elimination of ordinary garbage collection during the planning run.

The PFPS Inferencing Mechanism

The Symbolics implementation is based on a blackboard framework using a group of planning modules controlled by a higher-level program called the Logical Kernel. The planning modules deal with the various areas of expertise, such as retirement, investment, debt, education funding, and taxes.

The separation of knowledge sources into planning modules whose invocation is controlled by the Logical Kernel and the object-oriented database meets the definition of a blackboard model proposed in Engelmore, Morgan, and Nii (1988, pp. 1-22).

In this way, PFPS consists of the "many and varied sources of knowledge" described by Ed Feigenbaum in Engelmore, Morgan, and Nii (1988, p. v), each of which contributes to the formulation of a plan for a given client.

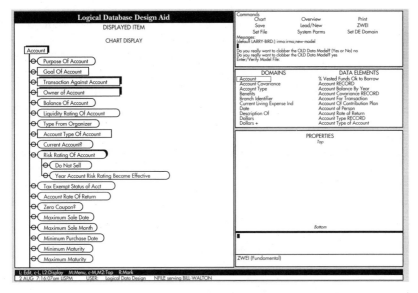

*Figure 4. Account Section of PFPS Database Structure
(rectangles represent data records, ovals represent data
elements of the record, shaded ovals represent subrecords)*

Each knowledge source contributes opportunistically to the formulation of a solution through the actions of the Logical Kernel.

The Logical Kernel, acting as a manager, traffic cop, and simulator of the passage of time, directs the year-by-year creation of recommendations, any backtracking that is required, and the arbitration of client goals. Backtracking is utilized by the planning modules to try various strategies in the year-by-year processing as well as to restart the plan in the event of overall goal failure.

In the first round of planning, if a financial goal cannot be satisfied in a given year by any means, the shortfall is recorded, and year-by-year processing continues. At the end of the planning horizon, the years and amounts of shortfalls are then analyzed to develop a revised set of goals for the next round of planning. This first round of goal arbitration is a major initial step toward the generation of a successful plan because it prunes away a large fraction of the possible search space. Of course, if all goals are satisfied on the first round, the planning ends successfully and a report file is created for uploading to the IBM.

On subsequent iterations of the year-by-year planning, goal failure in a year results in immediate goal arbitration (search-space pruning). This goal arbitration is less extensive than the first-round goal arbitration because the goal levels are now much closer to an achievable set of goals. Processing continues until a set of achievable goals is attained or until the overall level of achievement falls below a system threshold.

Many of the planning modules might be considered expert systems in and of themselves if they were separated from the rest of the system. Thus, for example, the funds-acquisition module (debt and other sources of funds) has knowledge of conventional lending rules regarding debt service–income ratios, different types of debt instruments (home equity, mortgage, unsecured demand loans, and so on), the tax status of various types of debt interest, IRA and defined contribution plan withdrawal regulations, and tax consequences. With this knowledge, in any year in which funds are required from sources other than normal income and assets, the funds-acquisition module searches through combinations of loans, seeking a low after-tax interest cost combination that satisfies typical liquidity rules; if necessary, retirement plan withdrawals are brought into play.

System Deployment

The personal financial planning service based on PFPS is currently being offered in Rochester, New York. The product has been available in this area since late 1987. The quality of the plan is comparable to many plans developed by existing traditional personal financial planning services and far surpasses any automated planning systems of which the authors are aware. Fees to a client for a PFPS-generated plan start as low as $600; other plans approaching the level of quality offered by PFPS can cost several thousand dollars. The plan itself is provided on a fee basis; implementation assistance is offered to the client, but this relationship is not required.

PFPS and the personal financial planning product were developed to fill a strategic need of the bank to provide affordable, high-quality planning to its broad base of customers. Although the stand-alone profitability of the operation is one key measure of success, it is being found that the relationships with customers begun through the planning process are key to the development of additional bank business. Future marketing plans are under review but could include wider distribution through the Chase Lincoln First market area and distribution through corporations and other institutions as an employee benefit.

Acknowledgments

The authors would like to thank Bruce Sather for his work on the design and implementation of the IBM side of PFPS, Eric Bush for his work on the original Prolog prototype, and Phillipe Brou for his work on the investment planning module. We would also like to acknowledge the work of Janet Harman and Patty Miller in the on-going system enhancements.

References

Curtice, R, and Jones, P. 1982. *Logical Database Design.* New York: Van Nostrand and Reinhold.

Engelmore, R. S.; Morgan, A. J.; and Nii, H. P. 1988. *Blackboard Systems.* Reading, Mass.: Addison-Wesley.

Moon, D. A. 1986. Object-Oriented Programming with Flavors. In Object-Oriented Programming Systems, Languages, and Applications (OOPSLA) '86 Conference Proceedings. *Sigplan Notices,* 21 (11), New York: Association for Computing Machinery.

Weinreb, D.; Feinberg, N.; Garson D.; and Lamb, C. 1988. An Object-Oriented Database System to Support an Integrated Environment. Unpublished paper.

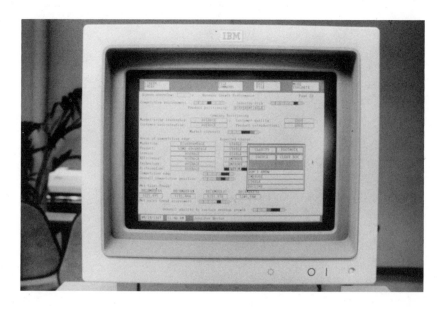

Syntel™: An Architecture for Financial Applications

Peter E. Hart

The 1980s can already be seen as the decade in which the expert system technology developed in research institutions during the 1970s was first exploited on a substantial commercial scale. Naturally, the initial applications of expert systems were guided by the lessons of the 1970s, and made as much use as possible of the familiar rule-based and frame-based representations that were first developed to address tasks in medical and other technically oriented domains. By heeding these lessons, exploiting established technology, and judiciously selecting applications, a significant number of expert systems were successfully fielded (Feigenbaum, Nii, and McCorduck 1988).

In our experience with financial expert systems during the past five years, we quickly found this strategy to be infeasible, chiefly because the existing technology was poorly matched to the demands of the selected financial applications. As the application requirements were better understood, our design response led to a new architecture that departs sharply from the rule- and frame-based architecture that dominates—and indeed is virtually synonymous with— expert systems today. The new architecture, termed an active functional system, is embodied in a programming system called Syntel™ that has been used as the basis for two commercial products: the Underwriting Advisor™ system and the Lending Advisor™ system. Technical details of Syntel have been reported elsewhere (Reboh and

Risch 1986; Duda et al. 1987; Risch et al. 1988); here, we focus on the application setting, describing only enough about Syntel to (hopefully) make the discussion intelligible.

Design Requirements for Financial Expert Systems

Many opportunities exist to apply expert system technology within the broadly defined financial field (Clifford, Jarke, and Lucas 1986; Friedman and Jain 1986; Dhar and Croker 1988; Leinweber 1988). The great diversity makes it difficult to define a useful set of system requirements that adequately characterizes all such financial applications, but as always, this difficulty is greatly eased by focusing on a selected subset.

Our own interest centered specifically on two applications: underwriting commercial insurance risks and assessing commercial credit worthiness. These two applications are similar in many respects. They are both concerned with estimating or assessing numeric and symbolic quantities (among which financial risk is an important but not unique instance); on investigation, both tasks prove to involve comparable styles of reasoning; both involve substantial amounts of case-specific, or transient data as well as permanent reference data; both serve business users with comparable attitudes and needs; and, although both applications are largely self-contained, both must exist in the real world of commercial data processing with its attendant interface and operational requirements.

Even this brief summary of application requirements suggests a number of technical implications for system architecture. We review here the most important of these implications, noting where appropriate why conventional rule- or frame-based approaches do not appear well-suited to the needs.

Assessment vs. Classification Reasoning Styles

A fundamental reasoning task for our applications is to assess (or estimate) the value of a continuous quantity such as projected profit. Any number of such subassessments—whether numeric or symbolic—are then weighed together to perform the overall apples-to-oranges assessments that are part of every business analysis (for example, in assessing the importance of a firm's poor financial performance against its attractive market and strong management team). This reasoning style contrasts with classification (or diagnostic) reasoning styles in which a selection is made from among a finite set of alternatives, and in which rule- and frame-based representations are so well-addressed.

The distinction between estimation and classification might superficially appear unimportant; one could imagine making a continuous variable discrete, thereby converting estimation to classification and making a rule-based approach an obvious one. Unfortunately, this becomes increasingly unattractive with increasing knowledge-base size because the number and complexity of rule antecedents needed to identify various combinations of assessments increases exponentially.

Inexact Reasoning

We expect the system to offer advice (more properly, assessments) in the face of imprecise knowledge or missing input data. Obviously, we prefer to have a smooth transition in the exactness of computed assessments as input data are made more complete. This requirement suggests the need for a principled approach for dealing with missing data, default assumptions, and imprecise knowledge. More specifically, because of the importance of continuous variables, it suggests using probability distributions to represent the several forms of inexactness.

User Initiative

We have observed that users insist on retaining the initiative in moving through the system. This observation suggests that the overall control scheme has to be data driven like the standard spreadsheet applications which are so familiar to business users. However, data-driven approaches can lead to serious inefficiencies when used on large knowledge bases; means for controlling excessive computation must be part of the basic design.

What-If's?: Controlling Side Effects

Users frequently engage in *what if?* explorations in which input data are repeatedly entered, modified, and then returned to their initial state. Understandably, users expect system-generated output to be consistent with restored input regardless of intervening computations. This requirement suggests the basic internal computations must be referentially transparent, so that restored output values depend only on the restored input values and not on sequence or side effects. We note in passing that applications built on conventional expert system approaches usually embody procedural escapes to the underlying implementation language, a practice which considerably complicates the problem of satisfying this requirement.

Data Management and Data Models

The number of variables needed to describe a single case (for example, a borrower) is large enough to require a significant data management facility. These transient data are accessed and modified so frequently that it is desirable for efficiency reasons to have the data management facility be an integral part of the expert system, rather than to rely on calls to an external database server. For theoretical and pragmatic reasons, it is most desirable to have a uniform data model for both transient data, permanent reference data, and expert knowledge.

The User Interface

Business users operate in a world of standardized business forms. To be acceptable, the user interface has to mimic this familiar world as closely as possible.

The Syntel Programming System

Now we sketch the principal elements of Syntel, drawing particular attention to the relation between architectural features and the application requirements noted previously.

Value Tables and Probability Distributions

Syntel uses a *value table* as its means for storing permanent reference data, case-specific input data, and the assessments derived from these data.. A value table can be regarded as a single database relation having an arbitrary number of keys and exactly one nonkey column. For example, the value table Revenue*(Year,State)* might hold historical and projected values of revenue for several years and states. Formally, value tables fit the functional data model (Gray 1984).

Each row, or *instance*, of a value table can contain in its nonkey column either an exact (numeric or symbolic) quantity, a probability distribution over this quantity, or a token indicating that the quantity is entirely unknown. Value tables are supported internally by efficient data structures and indexing schemes in recognition of the corresponding need to handle substantial amounts of transient data.

Primitive Functions

Syntel uses a collection of some 60 primitive functions to represent how input data are to be combined into progressively higher-level assessments. In addition to the usual classes of arithmetic, logical, and string-manipulation functions, special functions express how subassessments are to be weighed against each other to produce a higher-level assessment. Other functions modify the structure of value tables (for example, by reducing the number of keys).

Primitive Syntel functions are considerably more complicated than the functions found in conventional procedural languages. First, notice that a single function maps a set of value tables into a single value table; as a simple example, the function **Difference** could be used to map Revenue [*Year,State*] and Cost[*Year,State*] into Profit[*Year,State*]. All functions make use of a joinlike operation when computing derived instances. In addition, because function arguments can be probability distributions, primitive functions must map a set of distributions (assumed independent) into a derived distribution.

Primitive functions for knowledge representation mesh with value tables (note that value tables mathematically are extensional functions.) Taken together, this representation proved to be well matched to application requirements. Most importantly, it accommodates the reasoning style characteristic of the domain (including graceful degradation when data are absent); it supports what if's? because primitive functions are referentially transparent; and, with the control strategy described later, it provides the user-initiative capability that proved essential.

Forms: The User Interface

Syntel incorporates a major sublanguage called the *forms system* for defining the appearance, structure, and behavior of typical business forms. The role of the forms system extends much beyond being a conduit for input and output information. Indeed, it is an integral part of the overall control mechanism.

Data-Driven Control and Efficiency Issues

With a few exceptions, Syntel operates in a data-driven mode. Whenever a value table element is changed (whether by the user or by the system accessing external data through programmatic interfaces), Syntel re-computes functionally dependent values to maintain consistency with the new independent values. For this reason, we term Syntel an active functional system: Each function springs into action when its arguments change. As noted earlier, this architecture provides user flexibility at a potentially large computational cost.

Syntel employs several interacting mechanisms to control this cost. Using information derived from the current state of the user interface as well as from an extensive compile-time analysis of functional dependencies, the run-time system effectively implements a strategy that is easy to subscribe to but less easy to implement; this strategy is to identify the minimal set of logical dependencies, to compute only those dependencies which are required to support the current display state, and compute them only once.

User efficiency is at least as important as system efficiency. Syntel can conditionally display objects, a capability that allows an interactive session to be dynamically tailored to reflect the current state and, thereby, shield the user from irrelevant information or data requests.

Implementation Status

The Syntel formalism has been fully implemented since mid-1986, following three years of development, and supports a family of knowledge bases that define the Underwriting Advisor and Lending Advisor products. These products were originally developed on Xerox 11xx workstations. However, they are delivered to end users in a cooperative (or distributed) processing environment consisting of an IBM System/370 mainframe under MVS/XA using CICS and an IBM PC/AT or PS/2 workstation.

Each knowledge base contains several thousand value tables and about the same number of functions. A single value table might hold hundreds or even—in exceptional cases—over one hundred thousand instances. The depth of subassessments (more formally, the depth of function composition) can exceed 300. We know of no principled means for directly comparing functions with rules or frames, although for reasons described earlier it does seem clear that a much larger number of rules would be needed to represent equivalent expertise. Regardless of comparisons, these expert systems are, by any measure, large; careful attention to efficiency at all levels of design and implementation is mandatory.

Creating Syntel and all its surrounding software, as well as the various underwriting and lending knowledge bases, was not inexpensive; over five years and well over 100 person-years of direct technical development effort have been invested to date.

Deployment and Measures of Success

The Underwriting Advisor and Lending Advisor products based on Syntel have been purchased by insurance companies and banks in the United States and abroad. One might justifiably conclude that the innovative Syntel functional architecture has been successfully deployed, noting that: (1) the software system price ranges from half a million to several million dollars, (2) internal expenses of the purchasers make total installation cost rise above product price alone; and (3) each purchase was made only after a sophisticated institution conducted a lengthy evaluation and analysis. This conclusion is buttressed by the observation that the products go far beyond typical back-office automation; they directly affect the front-office decision-making process that lies at the core of every financial institution.

On a deeper level, evaluation is less simple. The roll-out schedule for conventional, large-scale system software in the financial industry is measured in calendar quarters or years, not weeks or months; expert systems are unlikely to shorten this timetable. Once fully deployed, a period of years is required before enough history accumulates to judge the consequences of the decision-making process (Did the loan eventually go bad? Did the building burn down?). Even then, it is difficult to establish objective evaluation standards because of ever-changing business conditions; for example, the changing business cycle, changes in tort law, changes in personnel, and changes in institutional policy.

For these reasons, a statistically valid analysis of results based on large, multiyear sample sets is currently beyond reach. Without this analysis, we are forced to rely on less formal evaluations based on our current users' experiences in processing actual lending and underwriting cases. On this nonstatistical basis, we have seen remarkably favorable results: Users at all levels of experience have reported many examples of "I would have overlooked (or misinterpreted) the significance of a critical piece of data regarding...." More importantly, system-generated assessments have been in full agreement with the judgments of senior credit and underwriting officers. At this juncture, it is fair to say that the issue of "Does the system work?" has been affirmatively resolved.

Summary and Conclusions

We have emphasized that the real-world requirements imposed on financial risk assessment systems differ in many important ways from what might be encountered in traditional application domains. The unique Syntel architecture, which arose as a design response to these requirements, has proven to be an effective foundation for building an unusual family of financial decision support systems.

Acknowledgments

The writer is merely the chronicler of, and contributor to, the efforts of a talented group of colleagues too numerous to name.

References

Clifford, J.; Jarke, M.; and Lucas, H. C. 1986. Designing Expert Systems in a Business Environment. In *Artificial Intelligence in Economics and Management,* ed. L. F. Pau, 221-232. Amsterdam: Elsevier.

Dhar, V., and Croker A. 1988. Knowledge-Based Decision Support in Business: Issues and a Solution. *IEEE Expert* 3(1): 53-62.

Duda, R. O.; Hart, P. E.; Reboh, R.; Reiter, J.; and Risch, T. 1987. Syntel: Using a Functional Language for Financial Risk Assessment. *IEEE Expert* 2(3): 18-31.

Friedman, J. Y., and Jain A. 1986. Framework for Prototyping Expert Systems for Financial Applications. In Proceedings of the National Conference on Artificial Intelligence, 969-975. Menlo Park, California: American Association for Artificial Intelligence.

Feigenbaum, E.; Nii, H. P.; and McCorduck, P. 1988. *The Rise of the Expert Company.* New York: Times Books.

Gray, P. 1984. *Logic, Algebra and Databases.* New York: Horwood and Wiley.

Leinweber, D. 1988. Knowledge-Based Systems for Financial Applications. *IEEE Expert* 3(3): 18-31.

Reboh, R., and Risch, T. 1986. Syntel: Knowledge Programming Using Functional Representations. In Proceedings of the National Conference on Artificial Intelligence, 1003-1007. Menlo Park, California: American Association for Artificial Intelligence.

Risch, T.; Reboh, R.; Hart, P. E.; and Duda, R. O. 1988. A Functional Approach to Integrating Database and Expert Systems. *Communications of the ACM* 31(12): 1424-1437.

TARA: An Intelligent Assistant for Foreign Traders

Elizabeth Byrnes, Thomas Campfield and Bruce Connor

Foreign exchange currency traders can not afford to think over multi-million dollar positions for long. In real time, they must examine large quantities of data; consider historical trends; determine what is relevant; and, many times in the course of a day, make the ever-critical decision to buy or sell. It is a high-risk, high-reward job of prediction where even the best traders are pleased with being right 60 percent of the time. At Manufacturers Hanover Trust (MHT) an expert system called the technical analysis and reasoning assistant (TARA), was built and deployed to assist foreign currency traders.

A Word about Trading

Currency trading is a form of price forecasting that deals with a deceivingly simple question: Where are prices going? Only three possible answers exist: up, down, or sideways. The success of a trader's forecast is based on the accuracy of analysis related to the timeliness of prediction. Large amounts of money can be lost or gained in instants.

Two popular approaches for determining price movement are fundamental and technical analyses. In *fundamental analysis,* the goal is to predict the supply and demand for a commodity, such as currencies. Fundamental analysts study economic forecasts, political events, and market psychology. *Technical analysis* is a forecasting method based on historical trend analysis. Technical analysts draw heavily on charting and statistical techniques. Neither approach has a standard formula for prediction, and both require substantial interpretation by a skilled trader.

Although some traders religiously pursue one strategy or another, most use a combination of technical and fundamental analyses. Because the technical approach is better documented, the initial deployment of TARA heavily favors this methodology. The knowledge base for the fundamental approach was developed, however, and is currently being tested. Like the wise trader, TARA will integrate both approaches in the future.

A Breakthrough for AI in Financial Services

Applying AI and expert system technology and techniques to the poorly understood and undocumented field of financial trading is viewed by many as risky and bold. Traditionally, expert systems have been successfully applied to domains characterized by recognized experts, bounded knowledge, adequate documentation, and small databases. In trading, the knowledge is fuzzy, no two experts seem to agree, and the volume of data needed in real time is large. The major breakthrough that TARA's success demonstrates is that expert systems can be applied to a much broader class of financial problems.

Why TARA Succeeded

TARA is successful for three reasons. First, the organizational structure for research and development was ideal. This project could be a text-book-case study for implementing new technologies. TARA had the support of top-level senior management who, from the beginning, committed trained AI people and a dedicated expert to the project. This expert, who was critical to the project's success, was selected by management for his outstanding performance and responsiveness to new ideas.

The second success factor involved the actual design of TARA. Because the domain was poorly understood, building TARA was a continuous, iterative process that involved hundreds of tactical decisions. Two questions continuously came up: From the vast amount of information available, what specifically does the trader need to know? Can we capture and represent this information on a computer in real time? Because few experts agree in this domain, it was critical to build around one expert but add flexibility to accommodate the preferences of many individuals.

Finally, the development team strongly believes that TARA could not have been implemented without a powerful AI development environment. The development tools provided by Symbolics and KEE™ greatly facilitated the iterative design process.

The Development Challenges

To build TARA, many technological and psychological hurdles had to be overcome. The biggest technological problem was integrating the live data feed with the technical models and the knowledge base without losing critical response time. The major psychological hurdles were gaining interest among traders with widely varying styles; holding this interest with a rich, yet logical and responsive user interface; and, finally, teaching traders how to use this higher-level tool to enhance their personal trading strategies.

TARA began as an experimental expert system project and quickly evolved into a powerful and sophisticated trader workstation. The decision to buy or sell could not be made without the technical models that identified historical trends and predicted major, minor, and intermediate changes in price. In addition, a mechanism for improving models by applying what-if analysis and examining past history was needed. This feature could not be implemented without a library of algorithms and an underlying database. Finally, unless the results could be quickly offered to the trader, the information was useless. Live data were critical, and a sophisticated user interface that supported both text and graphics had to be designed.

Deployment

The decision to explore AI's potential use on the trading floor was made in January 1987. After three months of interviewing the most successful traders and studying the domain, prototyping began. By June, a robust prototype was shown to traders and senior management, and the approval to proceed was given. An additional Symbolics machine was purchased, bringing the total number of computers to two.

Initially, an 18-month development schedule was planned, but after six months, pressure to test the viability of the system began to mount. Deployment was advanced to May 1988. When this move happened, the development emphasis shifted from adding intelligence to stabilizing the environment, perfecting the user interface, and connecting the live data feed. In addition, TARA began daily "paper trading" in preparation for the real money it would be handling.

In May, one of the Symbolics, equipped with a high resolution 19-inch color monitor, was moved to the foreign exchange trading floor.

In December, after eight consecutive profitable months, two additional machines were ordered. One of these machines is for foreign exchange, the other is for another trading area. In the case of the latter, new rules and features will be required, but the underlying analytic capabilities, access to live data, and user interface have immediate value.

Tara has evolved into a robust and powerful workstation. Typically, 50 windows are active (but not simultaneously displayed), and more will be added as additional prices are tracked. Fifteen years of historical data can be called into memory and accessed from any model running in one of these windows. For example, a six-month intraday model for Japanese Yen can require 300,000 data points. On the Symbolics, over 10 synchronized processes are either running or waiting to be run. The system is connected to a live, in-house data feed that is active 24 hours a day. Most of Tara is written in Lisp, but Intellicorp's KEE was used for inferencing and knowledge representation.

Throughout deployment, development and enhancements to Tara have been ongoing. The trading environment is dynamic and fiercely competitive. Tara must maintain its leading edge.

TARA

Tara is a stand-alone, real-time system that has four interdependent components: a live international data feed, a sophisticated graphic user interface, a model development and optimization program, and a knowledge base of technical and fundamental trading strategies. Currently, Tara tracks multiple currency, bond, and interest rates; updates over 30 different technical trading models; runs algorithms; and applies rules to interpret the models. The result is recommendations to buy, sell, or hold a market position.

Tara is designed to play two roles: skilled assistant and experienced colleague. As an assistant, Tara is efficient. With every price change, it evaluates whether to update any of the 30 or more active models. When an update is required, rules and algorithms are activated, and if appropriate, recommendations and alerts are displayed. All these events occur in a matter of seconds.

In figure 1, one of these models, a point and figure chart for the British pound, is shown. This technical trading technique filters out insignificant price movements and allows the trader to easily target the beginnings and ends of minor, intermediate and major trends. Time is not a variable; that is, a column can take a minute or a week to complete. Each box on a point and figure chart represents a trader-specified unit of price, in this case, 10 points. A column of Xs represents a steady increase in price; a column of Os indicates the reverse. Rules

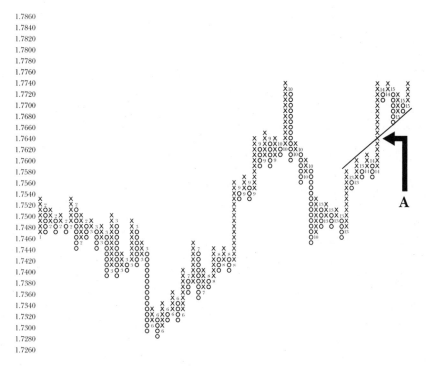

Figure 1. British Pound Point and Figure Chart

govern when to switch from one column to the next, the most common being the three-box reversal rule. This rule states that a new column can be started only when the price jumps three boxes, in this example, 30 points.

If TARA stopped here, it would not be much different from the existing technical charting programs available through vendors. What makes TARA unique is a sophisticated knowledge base of technical trading rules that helps traders decide which charts to use for a particular currency and how to interpret the patterns and trends. At this point, TARA begins to cross the line between assistant and colleague. The system builds sophisticated charts, but it also has knowledge about how to best use them in trading. For instance, suppose the system were tracking the point and figure chart in figure 1, and a new price was read, analyzed, and plotted at point A. A set of trend-analysis rules would fire, and a "buy British pound" recommendation would be given.

This example is certainly oversimplified. TARA's trading decisions are actually the result of multiple technical and fundamental factors. One model can be a good indicator in certain economic situations but terrible in others. Some models work well in tandem but are not useful alone. One of TARA's great strengths is the ability to perform this multifactored analysis in real time.

TARA's knowledge base does not yet include rules about the unforeseen business, economic, and political events that influence the market. This information usually travels by newswire or word of mouth. A trader can act on this information immediately, but TARA can not, at least for now. This broader, more complete market perspective is what the trader adds to TARA's analysis.

In sum, TARA adds objectivity, thoroughness, and speed to the decision-making process, but it does not replace traders. Instead, the system is a tool that enhances their abilities.

Development Costs

TARA's development costs are straightforward, consisting of the hardware and software mentioned earlier as well as the time of a four-person team. No outside consulting or resources were used beyond those typically provided by the software and hardware vendors. Since trading began, TARA has paid back all of its development costs as well as made an attractive return on investment.

Measuring Success

This project has two criteria for measuring its success: profitability and trader acceptance. TARA's technical analysis capabilities are extensive and proved profitable when tested historically. However, senior management wanted to see the system perform in live markets with real money. The profitability criteria also played a significant role in the traders' acceptance of TARA. After the first few months of profitability, traders began to regularly consult the system, either to request specific recommendations or to use the system's analytic capabilities to check hypotheses and test new models. By the end of 1988, requests for additional systems had been received.

How MHT Has Benefited

In addition to its contribution to bottom-line profitability, TARA has had a widespread effect on the organization's attitude toward technological innovation. Technology in banking typically meant back-office accounting, not front-office profit making.

For our trading business, TARA has given us a competitive edge in a rapidly changing business environment. Specifically, TARA's architecture provides a solid foundation for the trading floor of the future. Our sights are higher, and our goals are more tangible. TARA is not a hypothetical system; it is real and working.

Conclusion

TARA's benefits can be measured by more than dollar profits. TARA'S architecture provides a solid foundation for trader workstations of the future, where design is based on AI techniques integrated with state-of-the-art technology.

Many have questioned whether expert systems could ever work in a difficult domain such as financial trading. TARA calls for a positive reply.

Biotechnology

The PepPro Peptide Synthesis Expert System

Beckman Instruments, Genetech, Inc., and University of Houston

The PepPro™ Peptide Synthesis Expert System

*Philip R. Martz, Matt Heffron, Suresh Kalbag,
Douglas F. Dyckes, and Paul Voelker*

A *peptide* is a sequence of amino acids. A peptide can have anywhere from 2 to as many as 150 amino acids. At this point, the peptide is large enough to be called a protein. Twenty-one naturally occurring amino acids can appear in any order or combination in a peptide.

The ability to synthesize peptides is an important research tool in the biological sciences. Peptides perform a variety of roles under the more familiar names of hormones, inhibitors, growth factors, toxins, antibiotics, and proteins. This broad coverage encourages peptide research and drives the active synthesis of peptides for experimental use.

In the most common synthesis method, solid-phase peptide synthesis (Merrifield 1963), the scientist begins by anchoring the first amino acid of the sequence to a solid substrate. Then, with manual or automated (peptide synthesizer) chemistry, successive amino acids are coupled, one at a time, to the previous amino acid in the sequence until the peptide is complete. One can think of the synthesis as constructing a chain. A link in the chain must be opened; a new link (amino acid) must be inserted into the open position; and finally, the link must be closed. The synthesis must be structured so that at each step of the many links in the chain, only the link at the end of the chain is opened and added to.

Unfortunately, a synthesis is time consuming for the scientist, and worse, it might not go according to plan. The coupling of one amino acid to another takes roughly 90 minutes. Five amino acids can be coupled per day. A typical synthesis extends over several days. Several problems can occur during this time. The most likely is that an amino acid which should have coupled has instead coupled poorly or not at all.

If the scientist handles the coupling problems well, the synthesis is successful. If the scientist does not handle them well, the yield of the desired peptide is reduced, and the yield of undesired peptides is increased. The purification of the peptide, which is a major task in itself, can then become even more difficult. If the yield of the peptide is too low, the scientist might have to repeat the synthesis. Finally, economic losses occur in terms of time and materials.

We developed a program, the PepPro peptide synthesis expert system, that helps the scientist recognize and respond to the challenges of peptide synthesis. PepPro incorporates the knowledge of peptide synthesis experts. It uses this knowledge to analyze the peptide, predict coupling problems, and recommend appropriate procedures. PepPro produces a multipage synthesis report that summarizes the analysis and recommendations. The scientist can use this report to conduct the synthesis. One of PepPro's most powerful capabilities allows the scientist to write synthesis rules and add them to PepPro's knowledge base.

PepPro is a commercial product of Beckman Instruments, Inc. It is a 1.5 Mbyte program for IBM XT, AT, and PS/2 personal computers (PCs) and their compatibles. It includes a manual of 110 pages.

Several programs address the planning of chemical syntheses. However, these programs are general purpose for organic syntheses and do not specifically address the requirements of peptide synthesis. Two peptide sources that discuss solid-phase synthesis are Barany and Merrifield (1980) and Stewart and Young (1984).

Two of PepPro's functions are the subject of this paper. The synthesis function analyzes the peptide and provides the synthesis advice. The user rules function allows scientists to write synthesis rules based on their knowledge or experience and add these rules to PepPro's knowledge base. Two other functions, information and the user profile manager, provide important support roles to the expert system but are not discussed.

The Synthesis Function

The scientist makes almost all entries to PepPro by selecting commands with a mouse. To run the synthesis function, the scientist points

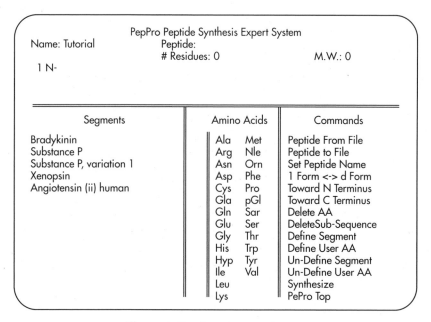

Figure 1. The Peptide Input Screen.

at the word synthesis on the PepPro top-level screen and clicks the left mouse button.

The first screen of the synthesis function is the peptide input screen (figure 1). This screen provides the scientist with a refined peptide-editing interface. It provides commands to edit amino acid sequences, save peptides by name, define new amino acids, and enter peptides from a file. Selecting any of these commands with the right-hand mouse button invokes the online help; PepPro then displays a statement of the command's function and operation.

Once the peptide is entered, the user selects synthesize. PepPro then begins a question-and-answer dialogue in which the scientist specifies the desired characteristics of the peptide and the synthesis. PepPro analyzes the peptide using a backward-chaining inference engine (MP) written by us.

PepPro's synthesis knowledge is encoded as if-then rules. A rule is shown in figure 2. The rule is used each time PepPro checks whether to acetylate an amino acid (a process that can reduce the formation of impurities) after it is coupled to the peptide fragment. The rule causes PepPro to determine if BULKY.BULKY = Yes. BULKY.BULKY is a parameter that indicates whether the coupling of one amino acid to another is likely to be slow. PepPro determines the value of BULKY.BULKY in another

Rule 1103: Rulegroup Peptide.Plan.Rules

If: 1. Bulky.bulky = Yes, and
      2. Either:
         3 A. (Pep.Charge - Current.charge)=1
         or
         B. (Pep.number / Frag.number)>.2

Then: Conclude Action is Acetylate

Figure 2. One of PepPro's Rules for Acetylation.

rule (not shown here) that examines both the sequence and an amino acids database. If the coupling is likely to be slow, the rule checks two conditions. If the Current.Charge of the peptide fragment is not equal to the Pep.Charge, the advice is to acetylate. The rule also succeeds if the number of amino acids in the fragment and the peptide differ by more than 20 percent.

PepPro's rules do not apply to the peptide as a whole. Instead, the rules apply to amino acids. PepPro uses its rules to analyze each amino acid separately in terms of its context of surrounding amino acids. It then determines the recommendation and makes an entry in the reports. PepPro repeats this cycle (evaluate the context, apply rules, write entry in report) for each amino acid in the sequence until all amino acids are checked.

PepPro produces three reports. First, the *synthesis report* is PepPro's analysis and recommendation of how to conduct the synthesis. The user can request that the report be produced in a concise or an expanded format. This report has three parts: general recommendations, a fragments list, and amino acid recommendations.

Second is the chemicals report, a list of the amino acids, reagents, solvents, and scavengers required to do the synthesis. This report is closely tied to the recommendations made in the synthesis report so that it reflects only those chemicals required for the synthesis in question.

The third report is the design input, a summary of the input from the question-and-answer dialogue.

The reports include commands on the right-hand side of the screen that allow for switching between reports (figure 3). A command saves the reports to a file. The change-answer command allows the user to change the requirements and have PepPro revaluate the synthesis. Commands along the last line of the screen allow for scrolling and paging through the reports.

PepPro Peptide Synthesis Expert System
Synthesis Report

14-Feb-1989 8:35:40

Peptide name: Test peptide

H- Asp Asn Cys His Gln Phe Nle Glu Ile Tyr
Asp Leu Lys Pro-OH

Synthesis
Chemicals
Design Inputs
Change Answer
Save Reports
PepPro Top

Difficulty (est.)	5 (range 1 to 10, 10 is difficult)
Relative Diff (est.)	9 (range 10 to 10, 10 is difficult)
Molecular weight:	1734.90
Isoelectric pt. (est.)	5.6
Resin:	1.50 g resin, Chloromethyl or PAM
Cleavage:	Low HF - High HF with DMS and Anisole scavenger
Theoretical yield:	1821.64 mg (1.05 millimoles)

Washg volumes of solvents (at 10 mL/g of resign) = 15 mL.
Coupling volumes of solvents (at 10 mL/g of resin) = 15 mL.

Figure 3. The General Recommendations of the Synthesis Report.

General Recommendations

This part of the synthesis report outlines the major characteristics of the synthesis (figure 3). The significance of the general recommendations section is that it establishes the primary considerations for conducting the synthesis: how to start the synthesis, how difficult it is, how to end the synthesis, and what techniques are required.

PepPro gives two difficulty ratings for the synthesis. These ratings are estimates on a scale of 1 to 10, with 10 indicating the most difficult. They are based on the peptide length plus rules that estimate how difficult it is to couple the amino acids in the peptide. Using these estimates, the scientist can plan an appropriate synthesis strategy by knowing how difficult a synthesis is likely to be. PepPro estimates the peptide's isoelectric point, information that is useful when the peptide is purified. It determines the appropriate resin linkage from the desired C-terminus form. The report shows the method for cleaving the peptide from the resin. It includes advice on appropriate chemical additives (scavengers). The report gives the theoretical yield based on 100-percent coupling efficiency and an estimate of actual yield that is determined by estimating the coupling efficiencies of the amino acids. Finally, the report gives the solvent and reagent (coupling) volumes that the scientist should use.

Fragments List

PepPro's account of peptide fragments (impurities) that are created during the synthesis is a significant capability (figure 4).

The report shows the number of residues, the fragment charge, and the source (either PepPro or user) of the rule that is responsible for producing the fragment. The fragments list helps the scientist monitor small but important details. The list can alert the scientist that given the current synthesis plan, a fragment similar to the desired peptide

	Charge	Residues	Source
Peptide:	-1	7	
Fragment:	-1	5	PepPro
Fragment:	0	4	PepPro
Fragment:	-1	3	PepPro
Fragment:	-1	2	User

Figure 4. PepPro's List of Peptide Fragments.

will be produced. The list can be used to determine a purification strategy. By examining the list, the scientist can decide whether purification based on size, charge, or a combination of the two is required to separate the fragments from the peptide.

Amino Acid Recommendations

The coupling of the amino acids is a crucial step in the synthesis and the synthesis activity most responsive to sound recommendations. Pep-Pro suggests a coupling strategy for each amino acid in the peptide. An example of an amino acid entry is shown in figure 5.

The recommendation gives PepPro's concerns, methods, and justifications for the coupling step. The first line of the entry shows that this advice concerns the coupling of Asp (Aspartic acid) to Arg (Arginine) preceded by Val (Valine). PepPro has selected a chemical group (OBzl) as a way to protect the amino acid throughout the synthesis. The advice indicates double-couple with Merrifield coupling for 15 minutes. PepPro provides a justification (Reason:) and any additional notes or suggestions (PepPro:).

When determining a recommendation, PepPro checks its rules as well as any rules that the user has written. Thus, the recommendation comes from PepPro's and the user's rules. PepPro marks recommendations that come from a user rule with an asterisk. In the figure, PepPro marked Merrifield coupling and 15 minutes with asterisks indicating that these recommendations came from a rule written by the user. Pep-Pro has also written an advise entry that indicates the recommendation PepPro would have made conflicts with the recommendation in the user's rules. The advise entry shows that PepPro would have recom-

#1	Boc-Asp(OBzl)	→	Arg-Val...Resin

Method: Double-couple. *Merrifield coupling
*15 min.

Reason: Double-coupling is advised due to the side groups of Asp and Arg.

Advise: HOBt, 30 min.

PepPro: Acetylating will produce a fragment with the same charge as the peptide.

Lab: Paul uses Merrifield for 15 min.

Figure 5. An Amino Acid Entry.

mended HOBt coupling for 30 minutes. The entry concludes with any notes (Lab:) that the scientist, by way of user rules, wrote for this step.

The User Rules Function

The user rules function is one of PepPro's most significant features because it gives PepPro unusual flexibility. We showed the central outcome of this flexibility in the previous section; by way of user-written rules, the scientist has a way to control the recommendations that appear in the synthesis report.

We wanted to provide user rules for three reasons. First, our experts agreed that they could not represent the range of experience that exists in the field. Second, a diversity of opinion exists on how syntheses should be conducted and which methods should be used. PepPro represents a significant approach but not the only one. User rules allows for this diversity of opinion. Third, peptide synthesis is a growing science; its rules are not hard and fast. Rapid increases in peptide synthesis knowledge could make PepPro obsolete. All these reasons indicated that PepPro should allow the scientist to personalize PepPro in terms of opinion and knowledge.

From the scientist's view, the reasons for writing rules are more immediate: to incorporate new knowledge, experience, or insights; to be more or less conservative than PepPro; to change one of PepPro's recommendations that is perceived wrong; and to add one's comments to a synthesis report.

Rule-Writing Mechanics

To run the user rules function, the scientist selects user rules from the PepPro top-level screen. The scientist can write rules for one of three rule groups depending on the type of synthesis. PepPro requests the name of the rule group that should be accessed and then displays the rule-writing screen (figure 6). This screen provides the commands necessary to write, edit, delete, and save rules.

The rule format is shown in figure 7. Using this format, the scientist can concisely express synthesis knowledge. On the one hand, the scientist can write rules that apply only when specific criteria are met. One example is shown in the first rule of figure 8, where the At Position and To: clauses establish a highly specific rule.

This rule applies only when Asp is coupled to the sequence Glu-Cys-Asp. On the other hand, the scientist can write completely general rules (second rule in figure 8) by using the designation Any AA. This rule always applies whenever any amino acid is coupled to any amino acid. The third rule demonstrates the use of rules as a sequence-dependent note-keeping device. The scientist is requesting that whenever PepPro encounters the pattern Gln-AnyAA-Pro at positions 2 through 8, it should enter the lab notes into the report. The fourth rule shows the use of the Coupling Other Than command. This command allows the user to state the amino acids to which the rule doesn't apply. This rule also shows one can write that something not be done (don't acetylate). The flexibility of PepPro's rule-writing scheme is noted by an interesting fact: The first rule in figure 8 is more specific, and the second rule is more general than any rule in PepPro's knowledge base.

User-written rules are not simply an adjunct to PepPro's rules. If desired, the scientist can write rules to establish a knowledge base that significantly enhances PepPro's knowledge base. The scientist can even replace PepPro's knowledge base by writing the most general rule premise, If Coupling AnyAA to AnyAA (second rule of figure 8). Because the user's rules are checked first and because this general rule always applies, PepPro's rules are never checked. With this rule in place, the scientist can write other rules for specific situations.

Integration with PepPro's Rules

A truism of rule-based systems is that each rule represents a unit of knowledge independent of other rules. In fact, this is rarely true. Rules interact in unexpected ways. Adding one new rule can easily disrupt an expert system, which raised the question of how to organize and use the scientist's rules and still maintain the integrity of PepPro's knowledge base.

We addressed this question in four ways. First, we keep PepPro's rules separate from the user's rules. Second, PepPro checks the user's rules before it checks its rules. If a user's rule applies, PepPro uses this rule and ignores its rule. Third, PepPro does not treat the user's rules equally. It ranks them from the most to the least specific by calculating

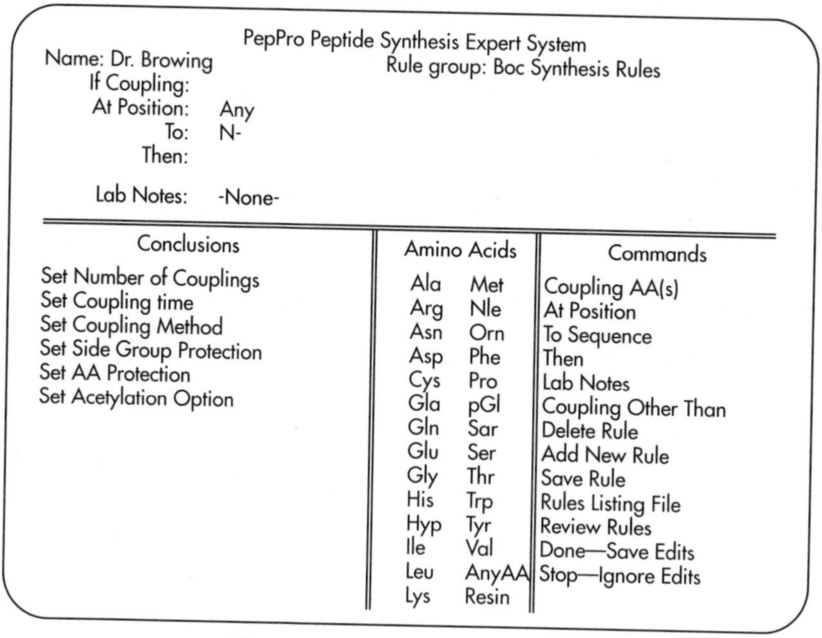

```
┌─────────────────────────────────────────────────────────────────────┐
│                    PepPro Peptide Synthesis Expert System             │
│   Name: Dr. Browing                  Rule group: Boc Synthesis Rules  │
│      If Coupling:                                                      │
│        At Position:    Any                                             │
│               To:      N-                                             │
│             Then:                                                     │
│                                                                       │
│      Lab Notes:     -None-                                            │
└─────────────────────────────────────────────────────────────────────┘
```

Conclusions	Amino Acids		Commands
Set Number of Couplings	Ala	Met	Coupling AA(s)
Set Coupling time	Arg	Nle	At Position
Set Coupling Method	Asn	Orn	To Sequence
Set Side Group Protection	Asp	Phe	Then
Set AA Protection	Cys	Pro	Lab Notes
Set Acetylation Option	Gla	pGl	Coupling Other Than
	Gln	Sar	Delete Rule
	Glu	Ser	Add New Rule
	Gly	Thr	Save Rule
	His	Trp	Rules Listing File
	Hyp	Tyr	Review Rules
	Ile	Val	Done—Save Edits
	Leu	AnyAA	Stop—Ignore Edits
	Lys	Resin	

Figure 6. The Rule-Writing Screen.

a score for every rule. For example, the more specific If coupling Leu to Asp is given priority over the less specific If coupling Leu to Any AA. When analyzing a peptide, PepPro checks the rules from the most to the least specific. This approach is a natural way to organize rules; the specific cases take precedence over the general case. Fourth, we do not prevent the scientist from writing unreasonable rules, primarily because we cannot confidently determine whether a rule is unreasonable. However, PepPro does use its rules to analyze the recommendations arising from the user's rules. PepPro's response to a rule that might be unreasonable is to place an advise entry in the synthesis report (figure 5).

Development

We started the PepPro project in late 1986. The development team consisted of one primary expert, three consulting experts, two knowledge engineers, and support personnel. PepPro's development included several versions of the software and a beta test. PepPro was used internally for about one year. It will become a commercial product in 1989.

PepPro was developed on Xerox 11xx series AI workstations and then compiled to run on PC. On the Xerox workstation, we used an

If Coupling:	*some AAs*
At Position:	*a range or list of positions*
To:	*a sequence of AAs*
Then:	*do the following...*
Lab Notes:	*for these reasons...*

Figure 7. The Rule Format for User-Written Rules.

expert system development environment (MP). MP was originally developed to support an earlier product, the SpinPro™ Ultracentrifugation Expert System (Martz, Heffron, and Griffith 1986). MP exists as Common Lisp source code with macros that conditionally expand, based on whether the inference engine is to run on PC or the workstation. It provides the developer with several capabilities that are not available in other PC-based shells:

- It allows the developer to build the expert system on a workstation (Xerox 11xx series), thereby taking advantage of the workstation's power and multiwindow bit-mapped interface. MP runs the expert system in a window that simulates a PC screen.
- It produces the completed system for the IBM PC. This capability is significant because Beckman's marketing studies indicated the IBM PC was the delivery vehicle of choice.
- It gave us control of the user interface and any specialized functions required by PepPro.
- It includes a sophisticated report writer that allows PepPro to write complex, multipage reports.
- It provides a collection of knowledge base development utilities, including editing rules and parameters, writing reports, noting inconsistencies, and finding redundant rules.

The initial development consists of building, reviewing, and editing the application in Xerox Common Lisp on the workstation using MP's tools and a simulated PC environment. A rule compiler then converts the rules into functions. Finally, the knowledge base is moved to the PC where it is compiled using Golden Common Lisp™.

Discussion

During the knowledge-acquisition phase of PepPro, the experts were required to formalize answers to previously unasked questions: How does one predict when to double-couple? How does one know when to acetylate? The answers to questions such as these evolved over time.

If Coupling:	Asp
At Position:	8, 9, and 15
To:	N-Glu-Cys-Asp
Then:	Double-couple active ester
	30 minutes Acetylate

If Coupling:	Any AA
At Position:	Any
To:	N-Any AA
Then:	Single-couple HOBt
	45 minutes Acetylate

If Coupling:	Gln
At Position:	2 through 8
To:	N-AnyAA-Pro
Lab Notes:	Gln-AnyAA-Pro cyclize here.
	Pause for coupling test.

If Coupling:	Other than Ile, Val, and Leu
At Position:	Any
To:	N-AnyAA
Then:	Symmetric anhydride 20 minutes.
	Don't Acetylate.

Figure 8. Examples of User-Written Rules.

One example is PepPro's acetylation strategy. Acetylation is a reaction that can be done immediately after a coupling step. If done thoughtfully, acetylation can produce capped sequences that improve the yield of the synthesis and make purification easier. Early on, several strategies were proposed and discussed. None of these provided a complete strategy. However, the eventual result was a complete and consistent strategy that evolved from the early strategies. Interestingly, this strategy (the knowledge) did not exist until the experts were required to formalize an approach.

PepPro has limitations. First, some limitations derive from the scientific community's incomplete understanding of the process of peptide synthesis. This incomplete understanding translates directly into limitations in PepPro. For example, the scientist can predict some coupling problems and can address these prior to the synthesis. However, other coupling problems can arise that were not predicted. The scientist is forced to respond with a countermeasure after the problem is detected. Second, the user rules function, although providing the structure for most of the rules a scientist might want to write, does not allow the scientist to write all the rules that apply to peptides. A third limitation is part criticism and part compliment. Some peptide chemists have noted that PepPro is good at finding the coupling problems they already know about but not good at finding problems they don't know about.

Despite these limitations, PepPro provides important benefits. The most important is the potential to improve the scientist's syntheses. PepPro should make these syntheses easier, less time consuming, and more efficient. A second major benefit of PepPro—and most expert systems—is the collection and formalization of the available knowledge on the subject. This formalization process improves the experts' understanding of their field, and consequently, it improves the expert system that results from their knowledge. A third major benefit of PepPro and other expert systems is that the collected information becomes knowledge. Information reported in the literature, although it might be the source for the rules in an expert system, is passive. It places the burden on the scientist to find, understand, and apply it. PepPro, however, actively responds to the problems posed to it. Finally, the user rules function enhances all the benefits mentioned earlier; it provides a growth path to improved syntheses, gives scientists a way to formalize and record their knowledge, and turns their synthesis notes into active knowledge.

Acknowledgments

For their contributions, the authors thank Phyllis Browning, Phil De Souza, Ed Faulk, Ed Fong, Manny Gordon, Jon Harbaugh, Michael Mokotoff, Jim Osborne, Ted Shapin, John Stewart, Bruce Wintrode, and Janis Young.

References

Barany, G., and Merrifield, R. B. 1980. Solid-phase Peptide Synthesis. In *The Peptides*. E. Gross and J. Meienhofer, Eds. New York: Academic Press.

Martz, P. R.; Heffron, M.; and Griffith, O. M. 1986. An Expert System for Optimizing Ultracentrifugation Runs. In *Artificial Intelligence Applications in Chemistry*. T. H. Pierce and B. A. Hohne, Eds., 297–311. Washington, DC: American Chemical Society.

Merrifield, R. B. 1963. Solid Phase Synthesis: The Synthesis of a Tetrapeptide. *Journal of the American Chemical Society*. 86:2149–2154.

Stewart, J. M., and Young, J. D. 1984. *Solid Phase Peptide Synthesis*. Rockford, Ill.: Pierce Chemical Co.

Emergency Services

A Knowledge-Based Emergency Control Advisory System
Australian Coal Industry Research Laboratories, Ltd.

MannTall: A Rescue Operations Assistant
Center for Industrial Research

A Knowledge-Based Emergency Control Advisory System for the Australian Coal Industry

John Aubrey and Zoltan Nemes-Nemeth

Following a major underground coal mine disaster in Central Queensland in 1986, the Australian government, through the then Department of Resources and Energy, announced that a special round of grants would be available specifically for coal mine safety. The aim of this injection of funds was to promote research and development (R&D) activity in an effort to help minimize both the effect and occurrence of such disasters.

Australian Coal Industry Research Laboratories Ltd. (ACIRL), in conjunction with a range of consultant subcontractors, put forward a research proposal to use expert system technology to develop an emergency control and advisory system for use in disaster management for underground coal mines.

The aim of the project was twofold: to develop an easy-to-use, approachable software tool that could guide, direct, and assist staff members in the administration and control of an underground mine disaster and to have this system available for use in training courses for underground rescue teams.

Work began on the project in July 1987, and the delivery of fully operating systems was made in June 1989. Prior to delivery, a series of prototype systems were configured and tested, involving the testing of several different inferencing mechanisms. A single implementation form evolved from these trials.

This final prototype uses a Digital VAXstation 2000, running VMS and using Nexpert as the inferencing engine. This configuration forms the main delivery vehicle for the system and contains the complete knowledge base to handle virtually any kind of underground mine emergency from initial response through cleanup and post mortem. Smaller subset systems using micro-computer hardware will be available for use at mine sites. These systems contain all the knowledge necessary to handle initial response actions for the mine sites.

Introduction

Three major problems exist during a mine emergency. The first problem is psychological: the need to remain calm and clearheaded in the face of potential loss of life or considerable financial loss. Complex statutory regulations must be remembered, and decisions must be made about how best to cope with the problems at hand. A system capable of presenting the correct information in the appropriate order would help ensure that all essential actions are undertaken.

The second problem is making sure that nothing dangerous is undertaken by inexperienced personnel and that time is not wasted on futile activities. A system capable of assessing the outcome of actions based on the status quo would address this need.

Finally, a need for coordination exists both between the various parties engaged in their diverse activities, and between the different work shifts as the emergency unfolds. A system that can track all the goings on and centralize information would help keep everyone informed.

The Emergency Control Advisory system (ECAS) is an attempt to supply these services to a mine site using knowledge-based technology. "It is not the purpose of ECAS to provide ultimate solutions to all emergencies, but to present what benefits may accrue from certain actions, what additional actions may be necessary to achieve those benefits, and finally, what dangers or hazards may be associated with such actions" (Hamment 1987).

In addition, the system must have certain key characteristics. It must be able to use available information to initiate and direct efficient problem solving. Information being entered into the system must be neatly managed and easy to access. The system should also be able to

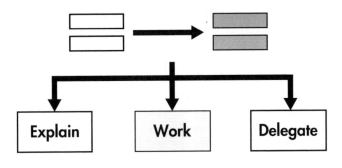

Figure 1

answer direct questions posed by the user and be capable of explaining and justifying its reasoning.

Representing the Problem

Procedures for handling mine emergencies can often be expressed as actions to be taken based on a series of preconditions. This condition makes the domain suitable for a production-rule system. Knowing which questions to ask is as important as being able to answer those questions. Some questions require evaluation after suggestive initial data are encountered, indicating data-driven inferencing. Other questions are heuristically activated, indicating metarules or some other strategy need to be employed. Direct queries require a classic, diagnostic, goal-driven approach. Obviously a mixed-mode inference engine is required to cater to these various requirements.

In ECAS, the rules serve three main functions (figure 1). First, they encode the knowledge in such a way that the inference engine can opportunistically work its way through a mine disaster problem. They make the system work.

Second, they hold the logic behind the solution and, through the explanation facility, allow justification of advice. Often, advice based on complex criteria such as the chemical behavior of gas mixtures and the vagaries of underground mine ventilation can appear counterintuitive. The explanation facility provides an essential service by demonstrating the logic behind apparently ridiculous suggestions. An example is feeding additional air over a mine fire. Of course, the fire burns with more vigor but in doing so does not leave incomplete combustion products, which are prone to explosion.

Finally in the case of ECAS, a need exists to be able to delegate inferencing to humans. In some cases, personnel might have to be out of the reach of the command center for periods of time. Rules that describe how to handle contingencies that might arise can be handed out in the form of advice as well as be a working part of the knowledge base. For instance, it might be necessary to remind miners searching for a suspected fire to regularly check the carbon monoxide (CO) level; if it goes over a certain value, then retreat to a safe distance, or use a breathing apparatus.

In terms of explanation, rules can indicate what use required data can be put to or what data are needed to work out a problem (figure 2). This facility is particularly necessary for the training aspect of the system. For instance, the detection of smoke issuing from a ventilation shaft can lead to an investigation of CO levels to determine whether an explosion has occurred; plus calls underground to see if any physical evidence can be found. Conversely, the same rule can indicate that one reason for taking CO readings would be to determine if smoke is coming from a ventilation opening. The trainee can be taught both warning signs and the use of a procedure. The same rule can be used in a number of ways, depending on how the application accesses the rule base.

A class system is used to differentiate objects on a functional basis and determine the firing priority of rules. Rules that warn about impending hazards must fire before those being used to make some general inference. This structure is necessary to ensure appropriate results when new rules are added to the system and to minimize the need to hard wire rules to promote required behavior.

The class structure can also be used to attach externally defined behaviors to objects, such as whether they will set up reminders or end up in special menus to draw attention to them. Objects can be defined in terms of the problems they address or the staff members who are most likely to be interested in them.

Disaster Management

ECAS is designed to be an appliance for dealing with mine disasters. It has an interface that can include data monitoring facilities common in a mining environment, such as gas meters. It allows the user to see only the parts of inferencing that are related to the problem at hand. As much as possible, the interface is designed to appear to be helping manage the disaster, with the expert system functions acting as an invisible driving force. The use of menus and icons is designed to make interaction with the system as simple as possible.

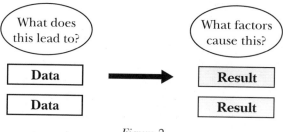

Figure 2.

On screen is the *Audit Log*, a time-stamped transcript of what dialogue occurred between ECAS and the user. Audit Log allows dumping to the file and printer with the click of a button, providing a permanent record of events.

Initially, most activity takes place through interspersed information-request menus and advice boxes put up by the system. Wherever possible, the user is given choices to click on with a mouse rather than empty space to type into.

A command menu allows the user to ask specific questions, volunteer or alter information. These tasks can be done whenever the system is asking a question or is idle. If the system is idle for a predetermined time, ECAS has the ability to dynamically display marked objects that can lead to further conclusions if instantiated. Alternately, the user can choose to ask specific questions or review progress.

A data-browsing submenu exists for seeing what is already in the system, for example, all information entered by users or conclusions arrived at by the system. An explanation submenu allows reasons for decisions to be explained; definitions to be given for key objects in the system; and by viewing the items queued for evaluation, the system's plan of attack to be previewed. Queries are usually not directed at general objects but at specific classes of objects depending on the type of query. ECAS itself manages the connections between the interface, the inferencing, and the external databases and subprocesses (figure 3). Some information, such as who to notify, automatically comes from the mine database, and other information, such as gas readings, can be sought from a number of sources. The user supplies any information that is not otherwise available.

Changing the Focus of Attention

In a conventional expert system there can be little chance to interrupt the inferencing process to evaluate new information. Questions are only asked when rules are being tested. Nexpert, the inferencing soft-

ware used in ECAS, moves one step away from this approach. It allows the user to randomly volunteer information. Its agenda structure means that new problems can be put on the "to do" list at any stage. Rule and object hierarchies can then determine which problem is addressed next. In the case of ECAS, it was thought that this approach might be too passive. One aim of the system is to ensure that tasks are not forgotten or overlooked. ECAS has the ability to cue crucial questions for reprompting at given intervals. The user is confronted with the question, and the consequences of any response can alter the agenda. This ability to que means that high-priority problems can be addressed as the need arises. Timer delays are kept in a metaslot attached to each object. They can be overwritten by the right-hand side of a rule if the need arises. Any object placed into the timer class is automatically reprompted after its delay time is exceeded.

Another problem arising from traditional inferencing is that answers which have been given can lose validity over time. The system needs to be able to know when conditions change, to be able to revaluate the situation. It is optimistic to think that staff members in the heat of a disaster will remember everything they once entered into the system. ECAS keeps a list of instantiated objects in the *if-change* class. These appear on a special menu as a reminder that these factors must be monitored.

During the initial stages of an emergency, it is important to enter initial information as quickly as possible and to reach some basic conclusions. It is not uncommon for certain advice to be entered as *not known*. Some of this not-known information might be necessary for crucial decisions, and ECAS highlights such objects in a separate menu as a reminder to find out as soon as possible. Information that might be useful but not crucial is also displayed, although less dramatically.

A third class of data exists that the system has the sense not to ask about because it is obvious the answer will not be available for some time. For example, the system might advise taking gas samples, and will need to know when they arrive. In this case,it cues the information in a when-known menu. It might or might not attach a reminder call to it.

Queries during Inferencing

ECAS is designed to be operated at either the novice or the expert level. To facilitate this arrangement, a number of objects are labeled to indicate rules are available to help determine them. An expert might know the response to a question immediately and directly attach a value to an object. A novice has the opportunity to reply not known, in which case the system backward chains in an attempt to find the answer. This technique means that trivial questions are asked only when required.

Some information requested by the system might require dangerous forays into the disaster area. For this reason, a why-ask facility was incorporated. This facility allows the user to see the use for the information and to determine whether it is worth pursuing. If the user decides not to find out, and the information is crucial, then it is still added to the needed-data menu to remind operators.

Whenever a rule is displayed for the purposes of explanation, it is possible to inquire about the way variables were determined or the outcomes of variables set by the rule. In many systems, it is not possible to trace an uninstantiated variable. In ECAS, it is possible to see any cause or outcome of a given variable whether it is instantiated. This format allows extrapolations from the why-ask or explain windows.

Knowledge Acquisition

Information for the system knowledge base was available from a number of sources:

- Statutory regulations indicate procedures that must be undertaken. They also indicate the correct delegation of authority and mandatory notifications. These regulations are consistent throughout an Australian state, although small details can vary depending on the physical characteristics of a mine, for example, how much methane is likely to be exuded from the local coal seam.

- Mine rescue techniques are the domain of district mine rescue stations. These techniques include a body of knowledge about rescue equipment, including gas analyzers and communications. Some individuals at each mine are usually trained at the local rescue station, but during an emergency, persons from throughout a district can be called in to assist.

- Laws of physics and chemistry play an important part in any disaster, particularly the behavior of gases and their potential to explode.

Although regulations can determine what needs to be done, the most effective way of implementing policy can require the personal expertise of a seasoned mine manager. In particular, this knowledge determines the order in which activities are most safely and effectively undertaken.

To collate this varied information, a retired mine manager and rescue station superintendent were employed to act as information coordinator and domain expert.

Implementation

It soon became apparent that the project needed to be divided into a number of components (figure 4). A small, initial action module need-

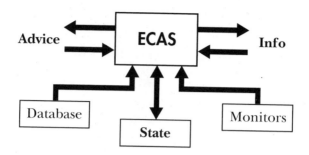

Figure 3.

ed to be available to virtually every mine site. The delivery system would have to be within the financial reach of small operations and had to be deliverable at a personal computer level. This system would have to be comprehensible to the most inexperienced of operators. Only a few choices would be offered, and data-browsing features would be minimal. It was anticipated that this module would consist mostly of a question-and-advice session driven by ECAS on the basis of incoming information. The management modules to be used by experienced personnel needed a far more powerful interface, capable of freely navigating the knowledge base. This module also needed the ability to manage the amount of information being accrued as the emergency proceeded. It was anticipated that this system would be used by the mine rescue team for its training as well as for deployment during an emergency. It was envisaged that the rescue team could be informed about the progress of effort following the disaster by the audit log, which could be transmitted by modem to the team before it set out. On arrival, team members could transfer the state of the working memory in the local system into their own version and continue management on the larger system.

Methodology

The original concept was to create a general purpose-knowledge base, then do knowledge queries on it. ECAS would translate user requirements into expert system and other requests and report back to the user (figure 5).

To implement the concept, it was necessary to create structures, much as a database has fields and relationships. Atoms correspond to different categories—advice, setting of states, information, and control

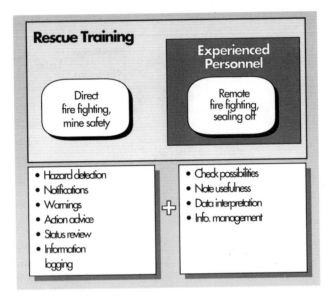

Figure 4.

features (figure 6). Obviously, the first task is to hide the control features from the user or have meaningful translations for them, such as "Look up the mine personnel database" instead of "DBload...."

The choice of expert system software was simplified because few products were capable of fulfilling the rigorous demands of the system. The software required both forward and backward chaining, device portability, and the ability to customize the interface. Nexpert offered all these capabilities, and its callable interface architecture also allowed us to customize the inferencing. Some parallel prototyping was done in Prolog to evaluate the usefulness of some of the anticipated functions. Although this practice functioned well on a small scale, it was thought that the C-based Nexpert would provide a much more comfortable and speedy development environment and would be easier to integrate with other mine-site software.

The inferencing procedure is an opportunistic mixture of forward and backward chaining (figure 7). Data-driven inferencing uses initial data to stack an agenda with useful rules. These rules are resolved, sometimes through backward chaining. As each of these rules fires or fails, it too adds to the agenda, thereby plotting a course through the knowledge base. Warning and informational rules are free to fire if they are triggered by the accumulating data. Object classes determine the priority of rules and dictate which will be fired first. The user has

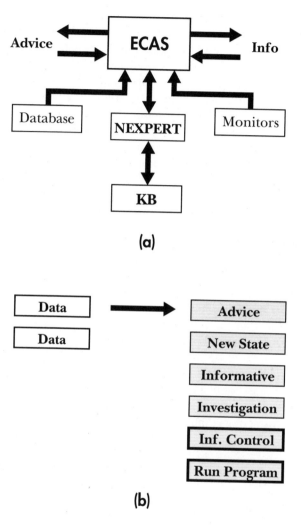

Figure 5 (a) & 6 (b).

the opportunity at any time to examine what is on the system's agenda by selecting an item from the explain submenu.

Initial Problems

The traditional expert system interface was the first problem encountered. The volunteer and suggest options available in the standard interface were daunting to the staff members who tested the initial prototype. This situation showed the necessity of creating a more intuitive interface that more closely modeled the way mine emergencies are handled. Explanations were supported only in the graphic network, which was difficult to operate, let alone comprehend.

Most of the menus offered were overloaded with options: The initial system had too many menu choices. People at the beginning of a problem were confronted with final solutions or, worse, control variables. Correcting this complication involved (1) partitioning the knowledge base to bring in rules and objects only when relevant and (2) establishing a classification system for objects, allowing them to be selectively displayed.

Rampant, disordered forward chaining caused advice to appear in a counterintuitive order. Originally, all rules had identical priorities and fired in forward-chaining mode in the order of writing. A framework had to be created using classes to allow the inference engine to fire important rules first.

Rules that had not-known values in them tended to get lost in the system. It was thought necessary to highlight such objects and try to get the required information and refire the rules when possible.

Because of the hypothesis system incorporated into Nexpert, no easy way existed to follow a forward-chained trail back to its origins in the rule network. A special browsing routine had to be written to facilitate this feature in the ECAS interface.

Explanations had to be parsed to make them more legible; even so, features put in to control inferencing needed to be hidden or effectively translated. Enough room had to be allowed so that some of the particularly verbose object names could be completely displayed.

Customizing ECAS

ECAS is a rule-driven system, and the questions it asks are generally required to satisfy a rule. To function properly, the system must have its queries answered. Where the answer comes from depends on the particular mine. On simple sites, most responses might have to come from the user. However, ECAS is designed to attempt to gain information from computerized sources if possible. With the order-of-sources metaslot, it is possible to specify a search path for data. External programs and databases can be searched for, and if unavailable, the system prompts the user. This means that the system functions similarly no matter how many peripheral information sources are available.

To function correctly in different locations and in mines of varying scale, ECAS must be able to cope with a range of mine-specific parameters. To accommodate this function, the system is designed to take ad-

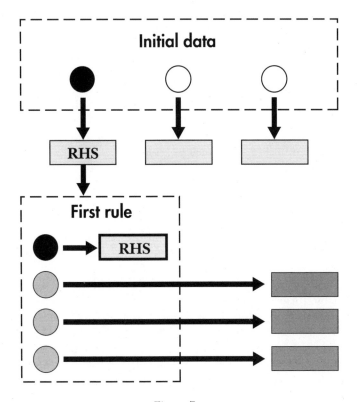

Figure 7.

vantage of a local mine database. The information in such a database can include the following:

- The physical characteristics of the mine, for example, layout and ventilation
- Local procedures for evacuations
- A catalog of facilities, for example, whether liquid nitrogen is readily available

The use of a database minimizes the need to alter the rulebase. In general, the system can decide whether some action is necessary and can leave the fine points of the procedure to the database.

Access to a database can speed up the system if questions normally presented to the user can be directly answered from the database. This organization can particularly aid the novice user, who might not have a good working knowledge of the mine site.

The type of hardware is also variable. Although the initial response system is designed to be universally available, some larger facilities might want a larger system that integrates with mine-site technology. From a practical point of view, the use of more powerful hardware allows more adventurous integration with other mine software, although the fine connections have to be customized to each individual software application.

The training system needs additional facilities to help in simulating predetermined emergencies and, perhaps, to allow the recording of scores for individual classes, and so on. However, the basic knowledge should be identical in each configuration.

Conclusions

All mine personnel who have seen the ECAS prototype in operation were favorably impressed and indicated they see a future for this type of expert system at the mine site. Mine rescue and emergency control is a knowledge-intensive domain where, fortunately, little chance exists to put training to the test. The ability to have this knowledge available in a usable form when the critical need arises is seen not only as utilitarian but also as reassuring. It is hoped that having a familiar system, which is used in training, also available during an emergency will mean that fewer rash errors are made initially and that experienced personnel will be free to devote their energies to complex activities, able to delegate the mechanical management duties to the system.

Acknowledgments

The authors acknowledge the following groups for their help and assistance with this project: the National Energy Research, Development, and Demonstration Council for funding this project; the University of Wollongong and Southern Mines Rescue Station for acting as sources of expert knowledge in the field of underground mine emergencies; Bellambi Coal Company for providing a place to test the ideas; Digital Equipment Corporation (Australia) Pty Ltd. for helping to develop the initial prototype; and the Australian underground coal mining industry for providing continued help, assistance, and enthusiasm.

Bibliography

Aubrey, J. and Nemes-Nemeth, Z. 1988. Developing an Emergency Control Advisor System. In Proceedings of the DECUS Australia Symposium.

Buchanan, B. G. and Shortcliffe, E. H. 1984. *Rule-Based Expert Systems.* Reading, Mass.: Addison-Wesley.

Englemore, R., and Morgan, T. 1988. *Blackboard Systems.* Reading, Mass.: Addison-Wesley.

Hayes-Roth, F.; Waterman, D. A.; and Lenat, D. B. 1983. *Building Expert Systems.* Reading, Mass.: Addison-Wesley.

Strang, J., and MacKenzie-Wood, P. 1985. *A Manual on Mines Rescue, Safety and Gas Detection.* Sydney: Weston and Co.

Harmon, P., and King, D. 1985. *Artificial Intelligence in Business.* New York: Wiley.

References

Hamment, J. 1987. Development of an Expert System for Hazard/Emergency Control and Training, Project Knowledge Acquisition Report #1.

MannTall: A Rescue Operations Assistant

Martin Brooks, Steinar Carlsen, and Atle Honne

MannTall is a stand-alone system for tracking personnel during an emergency on an offshore oil platform. Developed at the Center for Industrial Research in Oslo, Norway, MannTall is used at Saga Petroleum's Emergency Operations Room at Forus, on the West coast of Norway. The system provides a running analysis of the whereabouts of the platform crew in the form of upper and lower numeric bounds and the possible identities of crew members on the platform, in the sea, and on the various rescue craft. MannTall's input is a stream of messages regarding sightings and transfers; for example, "Abel, Berg, and one unidentified person have been picked up from the man-over-board boat by the helicopter." MannTall's analysis is used as decision support for the deployment of rescue craft. The name MannTall is Norwegian for census or roll call, literally meaning person-count.

MannTall is meant for use in accidents serious enough to warrant evacuation of the platform, such as an explosion or blowout. Such situations involve more than 100 people, the rapid mobilization of rescue craft, and a high volume of radio and telephone communications. Because of the large amount and the diversity of information involved, Saga's onshore crisis-management team faces an extremely complex decision-making task that it must perform under intense stress and time pressure. Its first priority is platform crew welfare.

Before deployment of MannTall in early 1988, personnel tracking was carried out by posting slips of paper, organized by people's names, on a bulletin board. Each slip contained information about a sighting or the transfer of named crew members. Several people were responsible for maintaining and interpreting the information. MannTall is a direct substitution for the bulletin board and its manual analysis. MannTall logs the information and automatically provides the analysis and is operated by one or two people.

MannTall has three advantages: The first is MannTall's ability to provide excellent best- and worst-case estimates in the presence of ambiguous and incomplete information, including reports about unidentified crew members. Formerly, estimates were based solely on intuition and limited by the complexity of the situation. The second advantage is that MannTall facilitates on-the-fly change of the persons responsible for personnel tracking. Because all message history and possible interpretations are on the computer screen, instead of in people's heads and their personal notes, the transfer of responsibility can occur efficiently and without loss of information. The third advantage is that fewer people are needed to carry out personnel tracking.

The value of MannTall's estimation ability is measured by the increased probability of saving a life. Test scenarios demonstrated that MannTall can discover the possibility of an unrescued person in the sea in situations too complex for intuitive human analysis. MannTall is successful in Saga's high-stress emergency operations environment because its analytic power is made available by way of a simple interface.

Problem Definition

During the first hours of an offshore emergency, sightings and transfers of people are reported to Saga's Emergency Operations Room. The information comes from a variety of sources, including the government's regional Rescue Coordination Center, the police, and companies employing people on the platform as well as the direct monitoring of offshore radio communications. Problems arise in interpreting the information because of the following:

- The same event might be reported several times.
- A message might be distorted when retransmitted through several channels.
- The contents of different messages might be conflicting.
- Some events might never get reported.
- Some messages refer to a single event, but others might refer to the results of several events.

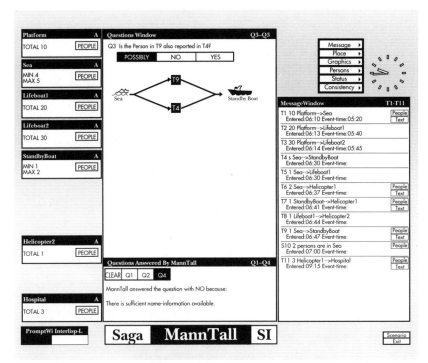

Figure 1. MannTall's Main Screen.

It is difficult for Saga's emergency operation staff members to discover all possible interpretations of the messages, maintain an overview of their combined effects, or know which additional information would reduce the uncertainty. MannTall performs these tasks.

MannTall's Capabilities

MannTall's initial state is a list of all persons on board the platform. (These data are maintained by an independent system connected to MannTall.) As the rescue operation evolves, messages regarding the transfer and current location of people are entered. MannTall computes upper and lower numeric bounds and the possible identities of the people at each place, including the platform, sea, and various rescue craft. Furthermore, MannTall generates questions to the user about the ambiguities inherent in the messages received so far. Answers to these questions tighten the bounds. The user enters the answers as the information comes in and, in the meantime, can use the questions as a what-if mechanism for exploring the possibilities.

MannTall's user interface was constructed with efficiency and clarity as the main objectives; both are necessary in the high-stress operational environment in which the system is used. All input operations are menu and mouse driven. The available information can be presented using several different viewpoints; it is always up to date.

Figure 1 shows MannTall's main screen layout. The message window on the right shows a summary of the input messages in chronological order. Each message describes a reported transfer of people from place

MannTall Brochure.

to place or the reported status of the people at a certain place. For example, the message labeled T4 reports two unidentified people being picked up from the sea by the standby boat. The message labeled T7 reports one person being transferred from the standby boat to a helicopter; in this case, the person's name is known and can be examined by clicking the mouse on the people button to the right of the message.

The small windows on the left of figure 1 correspond to places where people can be and give the maximum and minimum number at each place. The title bar of each window gives the place. For example, the *sea* window shows that there are four to five people in the sea. Clicking the mouse on the *people* button in this window pops up the list shown in figure 2. This list shows the names of the two people definitely known to be in the sea at the top, those possibly in the sea in the middle, and those definitely known to not be in the sea at the bottom.

The center question window, shown in figure 1, shows MannTall's questions to the user. The shown question asks whether the person reported in message T9 could be the same as one of the people reported in message T4. Note that T9 and T4 both report transfers from the sea to the standby boat, as shown in the drawing below the question. The answers to this question are *possibly, no,* and *yes,* presented as three buttons under the question. The answer shown is possibly; this uncertainty accounts for the range of four to five people in the sea and one to two

PLACE: Sea			
UNIDENTIFIED: MIN: 2 Max: 3			

DEFINITELY in Sea			(2)
Hedmyr	Isdahl		

POSSIBLY in Sea		(From 2 to 3 out of 64)	
Abeler	Haga	Nilsen	Urbye
Andersen	Hamran	Olufsen	Utgaard
Axelson	Ingstad	Oppedal	Utnes
Brooks	Isegg	Os	Val
Christensen	Jahre	Paasche	Vangen
Cox	Jerud	Poulsen	Velsvik
Dahl	Kaino	Quale	Waitz
Dignes	Kallevik	Qvigstad	Wechter
Eida	Larsen	Rammstad	Westli
Eik	Liberg	Randers	Xavier
Engan	Lie	Sand	Xiros
Fjeld	Lindoe	Sedal	Yhlen
Foss	Magnor	Seip	Young
Gjerde	Maltun	Thygesen	Zahl
Grape	Nettum	Tufte	Zeiner
Grave	Nguyen	Ulvin	Zoll

NOT in Sea			(4)
Abel	Berg	Carlsen	Prytz

Figure 2. Window Showing Who Can Be in the Sea.

people in the standby boat. The answer can be changed at any time by clicking on the *yes* or *no* buttons; new calculations are immediately performed. For example, clicking *no* gives four people in the sea (that is, min = max) and two people in the standby boat.

Below the question window is a window describing other questions that were automatically answered by MannTall. In this example, the user has clicked on Q4, and MannTall is explaining the basis for its automatically generated answer. The user can also pull up the text of question Q4 and change MannTall's answer if necessary.

MannTall is a relatively mature product; the computational core is augmented by a large amount of supporting functionality, including serial-line connection to Saga's persons-on-board database, status reports, information-logging facilities, consistency analysis, several forms of graphic presentation, and an extensive scripting facility for development and analysis of test scenarios.

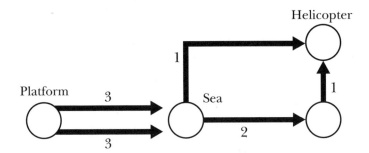

Figure 3. Two Types of Ambiguity in the Reported-Transfer Graph. There are 0 to 4 people in the sea, 1 person picked up from the sea in the standby boat, 1 to 2 people in the helicopter who have been in the sea.

Development and Fielding

MannTall was originally commissioned to demonstrate AI technology within Saga Petroleum. The safety area was chosen because it is open and contains no sensitive knowledge or information, and in this area, oil companies normally share their experience and results. Identifying the exact problem to be solved took several months.

We developed the MannTall prototype during 1986 in close cooperation with Saga. The initial work focused on the core functionality and user interface. As early as autumn 1986, experienced emergency preparedness managers at Saga recognized MannTall as a significant improvement over existing procedures.

In 1987, Saga began using MannTall as part of its regular simulated emergency safety exercises, validating MannTall's usefulness and usability. In mid-1987, Saga ordered an operational version, which was delivered in late 1987. The system was further refined during 1988 based on user experience.

Use during a major exercise in 1988 produced satisfying results: Throughout the entire exercise, the overview of the platform crew's whereabouts was as complete as possible given the available information, a situation seldom achieved before MannTall.

MannTall is installed in Saga's Emergency Operations Room and is part of Saga's formally defined procedures for reacting to an emergency. Fortunately, serious platform accidents are rare; as of spring 1989, MannTall was used only once in a real emergency. On January 20, 1989, the Treasure Saga platform in the North Sea was exposed to a potential blowout and was partially evacuated. MannTall was used continuously for 48 hours while people were ferried back and forth be-

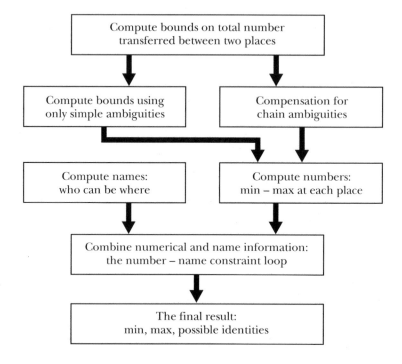

*Figure 4. Structure of MannTall's Heuristic
Counting and Estimation Techniques.*

tween the platform and other nearby platforms. Saga reports it was
satisfied with MannTall's performance.

MannTall was developed and deployed on the Xerox 1186. The majority of MannTall's four person-year programming effort went to the
original prototype and its functional extensions; three person-years of
nonprogramming effort went the into tests and evaluation and the integration into Saga's procedures. Development costs were about
$600,000. Atle Honne was the project manager from 1985 to 1986, and
Steinar Carlsen led the project from 1987 to 1988.

MannTall is generically designed so that it can be adapted to other
emergency organizations.

How MannTall Works

MannTall is a based on a reported-transfer graph having one node for
each place and one directed arc for each reported transfer of people
from place to place. Each reported transfer is labeled with pertinent in-

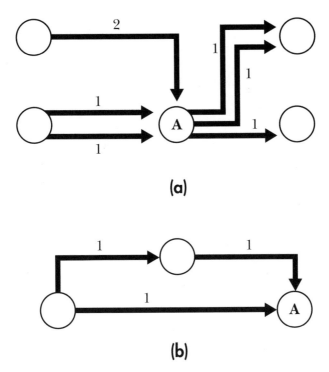

(a)

(b)

Figure 5. (a) Without Transfer Chains. The number of people at A is the difference between the totals coming in and those going out from other places; that is, ([2,2] + [1,2]) - ([1,1] + [1,2]) = [3,4] - [2,3] = [0,2].

Figure 5. (b) With Transfer Chains. One must be careful not to count the same person twice; the number of people at A is [1,2].

formation, including the time of the report, the number of unidentified people, and the names of those identified. MannTall recognizes certain types of ambiguities in the reported-transfer graph; these ambiguities are the basis for multiple interpretations of the reported transfers.

MannTall's core is the computation of minimum and maximum numbers—and possible identities—of the people at each place in the reported-transfer graph. In theory, one could find this information by creating one copy of the graph for each interpretation of the reported transfers; however, combinatorics make this impractical. Instead, MannTall keeps the ambiguity packaged in the single-graph representation and uses heuristic counting and estimation methods to interpret the reported transfers. This problem is novel and surprisingly difficult; its solution is the primary technical innovation in MannTall. The important property of these estimations is they bracket the true value; in particular, they never rule out an actual possibility.

MannTall recognizes reference ambiguities in the reported-transfer graph. For example, suppose there are two reports of three unidentified people falling into the sea from the platform. If these reports actually refer to the same people, then only three people are in the sea; however, if they refer to completely different people, then a total of six are in the sea, and if one person was common to both, then a total of five are in the sea. A reference ambiguity can also involve a chain of transfers; for example, suppose two men are reported picked up from the sea by the standby boat, then the standby boat reports the helicopter picking up one of the men, and the helicopter reports picking up a man who was in the sea. The third report might or might not refer to one of the men in the first two reports. Figure 3 shows the reported-transfer graph for these two examples.

MannTall uses a combination of techniques to compute the range of numbers and possible identities of the people at each place. The structure of the total solution is shown in Figure 4, where the arrows show the relations between subproblems. Some of the individual subproblems are briefly described in the following paragraphs; the point of these descriptions is only to give the reader a feel for the types of solutions.

Example No. 1

Compute the bounds on the total number transferred between two places.

Problem: Given a set of reported transfers, each from place A to place B, find the minimum and maximum total number of people transferred from A to B.

Note: The possibility of reported transfers referring to the same people is dependent on the user's answers to MannTall's questions.

Solution: A solution that ignores the names of identified people in the transfers is developed first. Tighter bounds are then derived by taking names into account. The solution has the form of a recursive counting procedure around a set of constraints.

Example No. 2

Compute the maximum and minimum numbers at each place.

Problem: Find the maximum and minimum number of people at a particular place.

Solution: The number of people at a place is figured as the difference between the number coming in and the number going out. Uncertain-

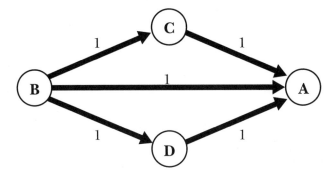

Figure 6. Overlapping Chains. The reported transfer from B to A is ambiguous with both the B-C-A chain and the B-D-A chain. The chain ambiguities cannot be analyzed separately because the person reported from B to A cannot simultaneously participate in both chains.

ty ranges—[min, max]—are added to get the totals in and out and subtracted to get the difference between the total in and the total out:

$$[min_1 , max_1] + [min_2, max_2] = [min_1 + min_2 , max_1 + max_2]$$
$$[min_{in} , max_{in}] - [min_{out}, max_{out}] = [min_{in} - max_{out}, max_{in} - min_{out}] .$$

If no chains exist, then this procedure is straightforward, as in figure 5a. With chain ambiguities, an additional technique is needed to recognize the possibility of counting the same person twice, as in figure 5b.

Example No. 3

Compensate for chain ambiguities.

Problem: Transfer chains create ambiguities where two people reported arriving at (or leaving from) the same place from (to) different places might in fact, be the same person. Given a network of chains ending at a place, compute the minimum and maximum number of people that could be counted more than once.

Solution: Overlapping and nested chains make the problem particularly difficult because it is not sufficient to analyze the chains individually, as illustrated in figure 6. The estimate is made by analyzing them separately while constructing a table coding the effects of overlap according to a topological classification of overlap types.

Example No. 4

Combine numeric and name information.

Problem: Numeric bounds on the number of people at a place are derived as described earlier. Lists of names of people definitely and possibly at the place are derived by other methods. These two estimates do not necessarily agree and need to be combined.

Solution: The numeric estimates and name data constrain each other according to a set of rules that are iterated forward until the results are stable. Each iteration tightens the bounds.

Acknowledgments

The authors wish to thank Birger Nordberg, Tor Stein Ølberg, Frode Leraand, and Bo Brännström at Saga Petroleum as well as Freddy Krogh and Morten Irgens at the Center for Industrial Research. The MannTall cartoon was drawn by Knut Westad.

The aim of the Department for Knowledge Based Systems at the Center for Industrial Research is to make AI technology operational in real-world problems. Our efforts are directed toward problems of high complexity within the industrial sector.

Law

An Expert System for Legal Consultation
Metropolitan Life Insurance Company

An Expert System for Legal Consultation

Paul Clancy, Gerald Hoenig, and Arnold Schmitt

Commercial real estate mortgage loans (loans secured by income-producing properties such as office buildings, apartment complexes, shopping centers, and hotels) are complex transactions. Metropolitan Life engages in many such transactions a year, and each transaction represents a substantial loan amount (sometimes as high as nine figures). The loan-closing process requires a significant amount of legal expertise and consultation during the commitment, documentation, and closing phases.

Although each transaction is unique, a general pattern exists to what is legally required, and prior transactions are used to create a framework for handling the current transaction. This framework building is time consuming and can result in an incomplete end product because the uniqueness of the current transaction precludes a perfect match with a prior transaction. A senior attorney ultimately has responsibility for the transaction, usually with assistance from a junior attorney or a paralegal. The senior attorney's time is costly, and it is desirable to minimize the time devoted to such a transaction.

An expert system was seen as a good way to provide help in identifying, organizing, and documenting legal requirements for commercial mortgage loan transactions. The main goals were as follows:

- To reduce the amount of senior attorney time spent on the transaction and the time for referrals among the senior attorney, attorney, and paralegal.
- To eliminate the need for extensive searches for precedent transactions and determinations of which are the most relevant to the current transaction (this elimination would contribute to reducing the time of the entire loan process, which would be beneficial for the legal process as well as the investment process).
- To provide accuracy, appropriate consistency, and reliability to the process through Metropolitan's nationwide commercial mortgage loan operation (the legal process is performed in six real estate investment offices spread across the country; invariably, some differences in the process occur because of individual approaches).

Although some interpretation of the law would be involved in establishing the knowledge base, the expert system would concentrate more on applying the established aspects of the legal process to the facts of each case. It was considered that this factual focus would provide the most benefit to the legal professionals at Metropolitan. This planned use of computerized help in Metropolitan's legal department was a significant departure from the department's typical computer use, which had primarily been for tasks such as word processing and time tracking.

The Solution

A team was formed of one knowledge engineer from Metropolitan's corporate expert system technologies unit and a senior attorney (Associate General Counsel) from Metropolitan's legal department. The attorney specializes in real estate transactions, has an interest in computer applications, and has served as chairman of an American Bar Association committee on the use of computers as a legal assistant. The team was given the task of building a reasonably robust, proof-of-concept expert system in three months. The proof-of-concept expert system would then be evaluated for acceptability; if given a positive evaluation, it would be expanded and distributed for use in all Metropolitan real estate legal offices. This proof-of-concept task was successfully completed, and the evaluation resulted in a decision (by the people in charge of the real estate investments section of the law department) to move forward with a production-level system.

To limit the project to a manageable scope and yet produce a practical working system, the problem domain was confined to the preparation of a detailed and comprehensive checklist. This checklist includes all the documents and other requirements to be satisfied to close the loan transaction based on the unique combination of facts presented by the transaction. The system considers various local, state, and federal laws and regulations as well as Metropolitan's lending practices.

Most of the knowledge acquisition was conducted through personal interviews with the expert. However, because the knowledge engineer and the expert worked in different parts of the country, some knowledge acquisition was conducted by telephone and mail. To accomplish the knowledge acquisition, as various checklist requirements and the transaction conditions they were contingent on were identified, different methods of knowledge representation were used to help review, expand, and define relationships. These methods included the use of pseudo-rules, lists of relevant facts and what requirements they affected, and lists of all requirements and which facts they were affected by.

Knowledge-acquisition sessions for the proof-of-concept task were held over a period of eight weeks, with an overlap during the latter part of this period for the start of knowledge base creation. Knowledge acquisition continued over another six months to reach a production-level system. During this expansion period, the knowledge base was also enhanced to reflect regional requirements.

The personal computer (PC) version of a commercially available expert system shell was used to create the knowledge base. The shell's internal report generator proved helpful in producing the printed checklist, which had to meet the extremely high standards appropriate to transactions of this nature (given the professionalism of the people and the amount of money involved.)

Another critical technical consideration concerned system memory. The system would be delivered on PCs with no more than 640 KBytes of random-access memory (RAM). The knowledge base grew quickly, and the total memory requirements for DOS, the shell software, and the knowledge base soon approached 900 KBytes. Thus, the knowledge base was designed to take advantage of the shell's provision to allow modular knowledge bases and its ability to move portions in and out of memory. This format made it possible to have the system run under the existing memory without any significant performance impact.

The initial (proof-of-concept) knowledge base consisted of 119 rules. When put into daily production, the knowledge base had expanded to 322 rules. Nine months into production, the system contained 428 rules, 351 parameters, 57 messages (each *message* is a unique screen format used to display different types of screen output to the user during a consultation), and 47 states (*states* are the high-level objects used to modularize the knowledge base; the system's rules, parameters, and messages are spread among the various states).

A System Consultation

A consultation with the checklist for income loan transactions (Clint) begins with a brief introduction and overview of the system for inexperienced users (which the user can skip). Clint operates by questioning the user, that is, an attorney or paralegal, about the various facts which must be known to identify the required documentation and closing conditions.

As an expert system, Clint asks the questions that the senior attorney responsible for the transaction would be posing to ensure the appropriate requirements are properly incorporated. The number of questions asked for each transaction generally ranges between 40 and 70, based on the responses to various questions. Accordingly, the system only pursues those lines of questioning and reasoning which are relevant to the transaction. Each user response results in (1) basic information that is necessary for the checklist which is displayed on the computer monitor or on a printed output, (2) a determination of certain documents or requirements that appear in the checklist, (3) additional questions to be posed to the user, or (4) any combination of these.

Examples of the types of factual information the user is asked to provide follow:

- The type of property for which the loan is being made: this information is needed to identify such legal concerns as franchise and management agreements, types of leasing requirements, and title policy requirements.

- The city and state in which the property is located: The system accounts for state and local legal requirements, which help determine such factors as building and zoning codes, environmental restrictions, and available title policy endorsements.

- The stage of construction of the project: Whether construction has not yet begun, is in process, or has been completed affects requirements such as agreements with construction lenders and the types of surveys.

- The organizational structure of the borrower, such as partnership, corporation, or land trust: This information is important for determining the specific types of documentation necessary to evidence the borrower's authority to accept the loan and execute the loan documents and in some states can be critical to determine whether the loan's interest rate is usurious.

- The type of interest that the borrower has in the mortgaged property, such as fee simple, leasehold, or beneficial interest in land trust: This information helps determine, for example, types of coverage required by the title insurance policy and the requirements for various legal opinions to be obtained.

The system allows unknown as an acceptable answer to some questions. However, these questions are relatively few because most answers should be readily available, or their omission would be too critical to allow. If an unknown answer is supplied by the user and accepted by the system, certain assumptions are made in determining the checklist requirements. Also, the output includes a list of all questions answered unknown, accompanied by a warning that the appropriate answers must be obtained and the consultation rerun to provide an updated checklist before closing.

Many other factors are also considered, such as type of loan interest (fixed or variable), the interest rate, proximity of the property to an airport, size of the property, liens, escrow agreements, and easements. Each combination of facts results in a unique set of documents and requirements that would not likely be found in a checklist for any one prior transaction and would not likely be within the knowledge of a user without substantial experience.

The system was also designed to check for inconsistent, invalid, and questionable responses from the user. When obviously inconsistent or invalid answers are entered (for example, amortization period is less than the loan term), the system warns the user and explains why the answers are inconsistent or invalid. The user is then given the choice of which responses to change and detailed instructions on how to make the changes. (For these situations, the consultation is not allowed to continue until the necessary corrections are made.) Clint also checks for questionable answers (for example, an unusually high interest rate or borrower's fee) and situations inconsistent with Metropolitan's normal lending practices (for example, a leasehold mortgage loan where the ground lease expires less than 20 years beyond the end of the amortization period). In these cases, the user is warned of the possible consequences, and asked for corrected information or confirmation that what was entered is indeed correct. This aspect of the system is important from a training standpoint and helps ensure correct input.

The output of the system, which is first displayed on the screen, is divided into several parts. First is a heading, which provides the basic transaction information such as loan amount and commitment expiration date. This information is followed by any appropriate warnings (for example, if any questionable answers were provided during the consultation and not changed when flagged by the system, the checklist contains strongly worded cautions about possible adverse legal or business consequences of the transaction).

The next section of the output contains the checklist items themselves. This section specifies the individual requirements that must be met before closing on the loan based on the information provided during the consultation (and those items which are universally appropriate and relevant to all cases). For ease of reference, this section is subdivided into six groups of requirements, as follows:

1. The loan documents group, which specifies the basic loan documents needed, such as promissory note, mortgage, tenant estoppel certificates, security agreements, assignments of rents and leases, Uniform Commercial Code (UCC) financing statement, Foreign Investment in Real Property Tax Act (FIRPTA) affidavit, and guaranty of loan documents.

2. The title and survey requirements group, which includes such items as title report and policy; zoning endorsements; survey of land; UCC search; as-build survey; appurtenant easement agreements; covenants, conditions, and restrictions; and considerations relating to flood zones and access to the mortgaged property.

3. The authority to execute documents group, which specifies the requirements to evidence the authority of the various parties to the loan (such as Metropolitan, borrower, and guarantor) to enter into the transaction and execute the appropriate documents.

4. The legal opinions group, which covers the required opinions of various attorneys such as Metropolitan's local counsel, borrower's counsel, and guarantor's counsel. These opinions deal with many areas, including the validity, enforceability, and due execution of the loan documents and the priority of liens and security interests.

5. The architectural matters group, which includes items such as approval of the plans and specifications by Metropolitan's architect, certificates from the borrower's architect, zoning permits, and certificates of occupancy.

6. The other requirements group, where anything not covered in the other five categories is included. This area runs the gamut from checks on usury laws, Employee Retirement Income Security Act of 1974 (ERISA) restrictions, relationship disclaimers and tax assessments to hazard insurance, finish and commission escrows, rent rolls, leases, and assignments of agreements.

Next the user is reminded to closely review the output to confirm that the requirements look correct and reasonable. If not, the system gives detailed instructions on how to change any information provided during the consultation (if the error was caused by incorrect input) or report any apparent inaccuracies in the output.

A reminder to store the consultation (to disk) is then given, with step-by-step instructions on how to do so. Most consultations are stored under the submission number or property name and the user's initials. The stored file provides a permanent record of the consultation for reference and auditing purposes. It also allows the consultation to be rerun at a later date when updated information can be provided without having to reenter all the information

After storing the consultation, the user is given the option of obtaining a hard-copy output in the form of a printed checklist. (Except for rare cases, this option would be exercised.) The printout is similar to the screen output, the basic difference being that it lists the requirements in columnar format. The first column shows each requirement, the second column has a sign-off space to document when each requirement is met, and the third column provides the opportunity for indicating that some requirements have been waived. (Note: Because some requirements are so important they can never be waived, the third column indicates *not waivable* for them, and does not provide a sign-off space for waiving.) An enhancement to the system will produce an alternate version of the printout, which will be provided to the borrower. This alternate version will not indicate which items are waivable, nor will it include any of the warnings, so that Metropolitan does not disclose its negotiating position to the borrower.

Next, if the user was not previously known to the system (Clint checks during the consultation), the system asks whether the user should be added to the list of known users. If so, a memorandum is printed containing the user's name, title, and other appropriate information. It's addressed (by the system) to the appropriate authority, and if signed by the associate general counsel in charge of the office (Clint plugs in the appropriate signature block), the name is added to the system.

Last, the system provides the option of printing a list of all questions and answers from the consultation. This list can be added to the loan file for additional reference if desired.

Conclusions

Clint was developed at a total cost of less than $75,000 (including the expert's time, the knowledge engineer's time, software for each office using the system, and travel expenses) and has been in use since April 1988. Because the system was designed to provide extensive automated help, its use was introduced in less than a day per office. The staff in each office was split into small groups to encourage an interactive training session, and each group was trained in a couple of hours. This training included not only actual system training, but also discussions on the rationale behind the conclusions reached by Clint in formulating the checklist. The system is currently running in three of the six

Metropolitan real estate law offices and will be in use in the remaining three offices by the end of 1989.

The online help facilities consist of prompts and explanatory material that guide a user through the process of using the expert system for a consultation, saving a consultation, changing a consultation, and printing a checklist. This online help (although always readily available, it is unobtrusive to experienced users) was seen as a key requirement to having the attorneys and paralegals quickly adapt to a new technology. Although printed documentation for the system was provided, it was more for background information and was not needed to use the system (even for inexperienced users, including some who had never before used a computer). To date, there have been few questions and few usage problems, but all were easily and quickly resolved.

Since installation as a production system, three enhancement releases have been issued. Assimilation of all three releases has gone smoothly using an install batch file to guide users through the installation.

Clint assists the less experienced attorney and paralegal by preparing a list of all documents and other requirements for a specific commercial real estate mortgage loan transaction. The list is detailed and comprehensive and can be as long as 15 pages. Many of these requirements would not be known by a person without substantial experience. However, the system is only an adviser; ultimately, the attorney who is responsible for the transaction has to be responsible for all the requirements in the checklist. Great care was taken in acquiring the knowledge for Clint, and one of the advantages of the system is the ease of updating it. It can be improved and enhanced any time a lawyer or paralegal comes up with an additional requirement that does not appear on a checklist and which the expert concurs ought to be added to the system. The expert then advises the knowledge engineer of the new requirement and the conditions under which it should appear on a checklist.

The expert system provides the following benefits:

- Time savings result in the form of faster identification of the legal requirements, the magnitude is minutes versus days.

- The system allows the checklist to be done much earlier in the overall loan transaction. Thus, the work necessary to satisfy the various loan requirements can begin sooner, and, in many cases, should enable Metropolitan to close loans at an earlier date.

- The ability of the system to save consultations makes the process of modifying a checklist because of an investment-dictated change much easier.

- The system provides company-wide accuracy and consistency in the identification of the legal requirements. As the process is handled in several dispersed offices, the system provides a common thread among the offices, and at the same time allows for some differences relevant to the specific geographic regions of the country.

- Senior attorneys minimize their time on the requirement-establishing phase of the transaction.

- The creation of a checklist can be performed by those too inexperienced to have produced it before. The system identifies the esoteric aspects of a case that formerly might not have been ascertainable to them. Of course, this feature enhances the training of new attorneys and paralegals.

- Use of the system by an inexperienced lawyer or paralegal has training implications beyond the specific transaction being worked on because it helps broaden their knowledge of all commercial mortgage loans.

- Allows for more timely and uniform response to business policy and statutory changes. A change can be introduced into the system and be in consistent use in all offices in a short period of time.

The Future

Clint has been so well accepted that consideration is being given to expanding the system to provide assistance during the commitment phase. The commitment takes place during the earlier phases of the loan process and is an agreement entered into between Metropolitan and the borrower, whereby Metropolitan agrees to make a specified loan under certain conditions. Because Clint identifies all the requirements for making the loan, it can also be used to check on the adequacy of the commitment to make certain that necessary or desirable conditions are not omitted. The system might also eventually be expanded to prepare initial drafts of the commitment and other loan documents. Additionally, consideration is being given to expanding the system to handle farm and ranch mortgage loans.

Finally, Clint might be made available outside Metropolitan. The checklist process is basically generic to all commercial real estate mortgage loans and is performed by a multitude of organizations, including law firms that provide contracted services to Metropolitan and other lenders as well as other lenders themselves (including banks and insurance companies). To summarize Clint's future, we envision it being used for other parts of the commercial real estate mortgage loan process; other types of loans within Metropolitan; and, possibly, other organizations.

Manufacturing Assembly

CAN BUILD: A State-of-the-Need
Inventory Simulation Tool
Digital Equipment Corporation

Charley: An Expert System for
Diagnostics of Manufacturing Equipment
General Motors Corporation

Harnessing Detailed Assembly
Process Knowledge with CASE
Boeing Computer Services

Knowledge-Based Statistical Process Control
Alcoa Laboratories, BBN Laboratories, University of Pittsburgh

Logistics Management System
International Business Machines, Inc.

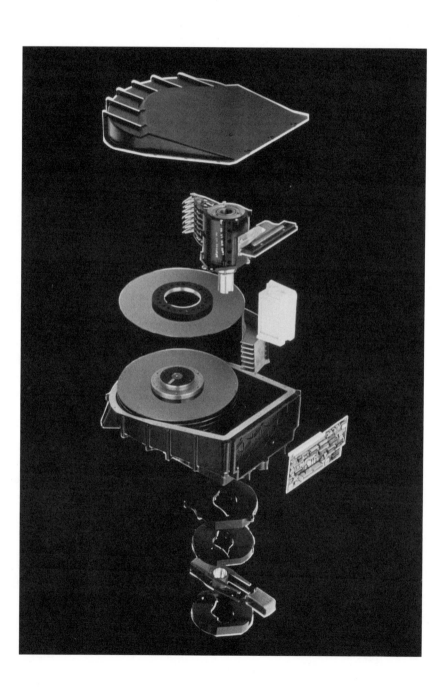

CAN BUILD: A State-of-the-Need Inventory Simulation Tool

Robin M. Krumholz

In any industry where the product line is constantly evolving, the accumulation of inactive (slow moving and obsolete) inventory can be a materials management nightmare. As a product declines toward end of life, decisions must be made about adding material and labor to build it up, disassembling it (salvaging usable materials from obsolete parts), and selling back unused parts to other vendors—or writing it off. Understanding these complex trade-offs is extremely difficult.

In 1985, Digital created a task force, the Inventory Programs Team (IPT), made up of experts from the materials, finance, and marketing groups. IPT's three-year mission was to find out what the company could do to reduce millions of dollars of inactive inventory. Its goal was twofold: to reduce inactive inventory to a minimum and to design a process to maintain inventory at these levels.

After a year of investigation and experimentation, IPT realized it needed more sophisticated tools than quarterly aging reports and spreadsheet applications to achieve its goal. The process it had developed was time consuming, subjective, and inefficient. It limited its efforts to a handful of products each quarter.

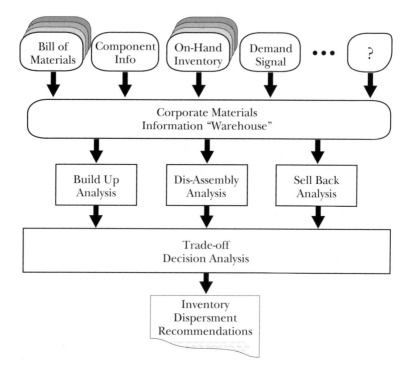

Figure 1. Model of Comprehensive Vision.

At this time, Digital's Applied Expert Systems Group (AESG) offered to assist the Digital Materials community by working with IPT to capture some of the IPT decision processes in a rule-based model. The joint AESG-IPT project began in August 1986. AESG assigned a knowledge engineer to understand IPT's business by interviewing members of IPT and studying IPT's goals and methods. Together, they drew up a comprehensive vision to model the trade-off decision process that IPT had developed. Figure 1 shows the model of this original vision.

The comprehensive vision included several different application modules that would share a common data environment. In the vision, each module was to analyze a different alternative for the dispersal of inventory. Analysis results were to be used as input to the final trade-off module. The trade-off module would use reasoning to select the alternative that best satisfied various sets of business objectives.

In addition to a significant development effort, implementing this vision required the creation of new data elements and data standards

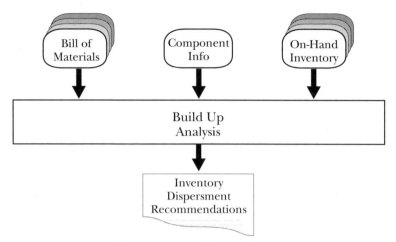

Figure 2. Scope of Phase 1 Project.

that would have to bridge separate divisions of the corporation. AESG's experience and development strategy (Gill 1987) suggested that the initial goal be narrowed in scope and made attainable within one year. The first phase of the project was intended to test the feasibility of the full model. At this point, AESG identified the piece of the model which would have the greatest impact on IPT's productivity and which would use only existing data. The original CAN BUILD system was the result of the focus on this narrowed model (figure 2).

Functionality

Can-build analysis is the process of simulating the financial impact of building various quantities of a product. The CAN BUILD system is an interactive simulation tool that allows a materials analyst to play what-if games with alternative build plans. Detailed inventory data are supplied to the system in a monthly "snapshot" (extract) from each of the scattered stockrooms and company inventory holders. Product and corporate reference information is supplied by data extracts from a number of corporate data systems. Bringing all this information together into a single system creates a decision support environment for management that has opened the door to a better, simpler way of managing inventory.

For any given product, the system identifies the total inventory dollars on hand that could be used to build a given product. It separates these dollars into inventory that is unique to the product (which can be used only for this product) and common parts (inventory that could be used for other products as well). The analysis can be done with inventory across the whole corporation, or it can use some geographic or functional subset of the inventory. The user can select a series of build quantities for simulation. For each build quantity, the system determines and tracks the inventory consumed, the additional labor and materials required, and the materials remaining.

CAN BUILD precalculates a set of recommended build quantities for the selected product, including the quantities at which significant business milestones would be reached. The system builds these numbers into its knowledge base. By looking at these recommendations, users gain insight into which build quantities they might want to simulate.

Through simple constraint-optimization rules, the system can model the natural manufacturing *food chain* (all the plant sources that feed each other during the manufacturing process, from raw materials to finished goods). In addition, it draws inventory from the minimum number of stockrooms to build any given quantity of the product. At any point, the user can request detailed information about the inventory used for the simulated build and the inventory remaining (the candidates for write-off).

CAN BUILD differs from other materials systems in that its purpose is strategic and tactical. Before this tool was developed, materials applications that had access to detailed inventory data modeled only what was within the four walls of any single plant. Corporate evaluations and decisions were almost always based on *rolled up* (summarized) information, usually provided in hard-copy reports and manipulated with spreadsheets. Management reporting was limited to what could be extracted from systems with an operational focus. The Sloan School of Management at the Massachusetts Institute of Technology refers to this approach as the *by-product technique*.

Before this project, it was not considered necessary to use corporate-wide stock status data for online interactive query. From its conception, the prime focus of CAN BUILD was to address management's strategic concerns rather than the daily operational issues of any one plant or group of plants. (Daily operations are a different and equally legitimate business concern.) By demonstrating to management the benefits of pulling information from multiple sources into a single data warehouse, we created the platform for this application and other envisioned decision support tools.

CAN BUILD in no way duplicates or replaces the systems that are used to manage materials within the individual plant. It provides its users with the capability of looking beyond their four-wall plant perspective, simulating various business alternatives, and seeing the impact of their decisions on all the plants above and below them in the manufacturing food chain.

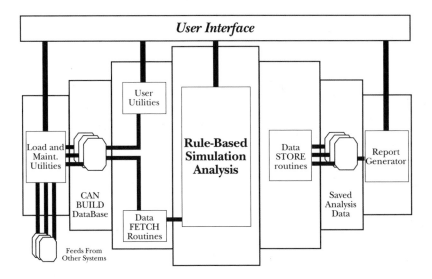

Figure 3. CAN BUILD *System Architecture.*

System Design

This application exemplifies the power of integration on two levels. First, it demonstrates the integration of multiple systems sharing information over a network without human intervention. Second, it also illustrates the integration of multiple software technologies, using the appropriate software techniques for each piece of the problem, to achieve a complete business solution. One can think of this application as a multiparadigm approach, where traditional data processing is one of the paradigms. We prefer to describe CAN BUILD as an integrated software solution, rather than an AI or expert system, because the rule-based part of the system is only one of several elements contributing to its success. The VAX OPS5 language interface provided the flexibility to use other VMS languages and tools (Cooper-Wogrin 1988); the developers were, thus, able to choose the most appropriate methodology and tools for each function.

The system architecture is a layered modular design, which separates the data-capture functions from data access and data manipulation. All data, including the results of any simulation analyses, are maintained in traditional data files that can also be shared by other applications. Figure 3 shows how the architecture is layered and how the rule-based elements of the system are isolated from the data environment.

The large central box in figure 3 represents the core of system rule-based modules. On either side of the central box is the interface layer, which passes information in and out between the VAX OPS5 environment and the data layer. The use of data layers gave the developers the flexibility to change external data extracts (in the outermost layer) and refine the rule-based modules without adversely affecting the other parts. Data layers also allow users and external applications to share the same data. The last element is the report generator. Although our solution has report-generating capabilities, users can use any fourth generation language reporting tool for customized output. The system is bound together by a menu-driven user interface designed by the users; it is extremely easy for any materials analyst to use.

AI and the Triangle of Change

When one looks at the quantity of rule-based code that was used in the baseline implementation of CAN BUILD relative to the quantity of traditional software tools, one might indeed question whether this application should be labeled an expert system. Although the initial prototype, consisting of some 200 rules, was written entirely in VAX OPS5, only 20 percent of the system (approximately 300 rules) was coded in a rule-based language in the baseline implementation. Less than half of these 300 rules (about 10 percent of the system) could be considered as doing any form of reasoning.

Initially, these statistics were disturbing to those who wanted to call CAN BUILD an AI solution. However, the percentages fit well in the triangle of change model developed by Dr. Gregory Gill at the University of New Hampshire (figure 4). As knowledge engineers, AESG members have been using this model to better influence and manage the organizational change associated with the development and use of expert systems within Digital. The triangle of change model shows that to successfully effect change, the affected organizations must value making investments in working better and differently (Gill 1987).

According to this model, only 70 to 85 percent of the resources in a business entity should be dedicated to the daily work necessary to getting the product out. The remaining 15 to 30 percent of resources is broken into two parts, with 10 to 20 percent improving the current processes and only 5 to 10 percent representing real change.

We use this model to show our sponsors that if they spend too much time getting day-to-day work done, little room is left for change. Therefore, they must respect the time required for projects that will help bring about change and commit themselves to investing people and time in the project.

However, the model is also intended to show that if the investment in changing the current work is too great, it can distract from getting

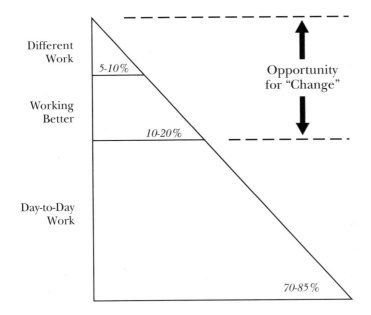

Figure 4. The Triangle of Change Model.

the day-to-day work done and lower business productivity. More important, change is likely to be more effective if it is evolutionary. Too much change too soon might be rejected. When we looked at the use of AI methodologies in CAN BUILD as a business solution, with the triangle of change in mind, we saw a direct connection to the model's implied recommendations for evolutionary change. We propose that having limited change (AI) helped build the credibility of the CAN BUILD system with the materials analysts and made it more acceptable to the materials business.

Development Strategy

CAN BUILD was developed through a participative design approach, using frequent interviews to acquire the necessary knowledge and iterative prototyping to define the tool. The project proposal—a simple, nine-page document—was accepted by the IPT sponsors in September 1986. The proposal included the original model; a description of the AESG development strategy; a brief description of the business needs that had to be met; and most important, a definition of success agreed to by all those involved. The description of the business needs included a list of 10 materials management questions for effectively managing inventory. These questions were difficult or impossible to answer using the existing tools and processes; at the time, the project team had only a vague idea about how to find the answers. Over the next six months, the developers would help the rest of the team experiment with different ways to arrive at the answers.

The definition of success was the result of a brainstorming session on the following questions: "If we are successful, how will we know it? What will be different two years from now?" IPT and AESG put together a list of business goals as criteria. Part of our strategy was to focus on the critical success factors of the business (Rockhart and Bullen 1986). The business goals that the team members laid out in this nine-page project plan helped us all stay grounded and keep our common goals in mind as we developed the functionality of the tool. This document remained the only formal documentation on the system until we implemented it. A part of our strategy was to also bypass the need for a formal functional specification. The prototype would serve this purpose.

The first prototype was developed entirely in VAX OPS5, using artificial data, in just seven weeks. This prototype captured the essence of the desired functionality. When it was demonstrated to the project sponsors, they gave their approval to get the tool working with real data. It served as a functional specification for subsequent efforts.

A second developer was brought onto the project. It took six other iterations of the design over the next four months before the team developed an appropriate knowledge representation of the problem, the baseline functionality was agreed on, and the developers began their final implementation design.

At the start of the project, the team had only a vague understanding of all the functionality required by the comprehensive vision. Experimenting with flexible data structures in a rule-based paradigm was invaluable. With each new design, the need for traditional data processing grew. As the functionality became well defined and stabilized, the developers wrote more and more of the solution in VAX BASIC, focusing the VAX OPS5 environment on those aspects of the system where rules provided an advantage.

The baseline system was implemented in September 1987, meeting the one-year goal. During the remainder of 1987, IPT used the tool to analyze some 15 end-of-life products. By identifying opportunities for revenue from inventory that would otherwise have been written off, IPT recouped for the company more than 30 times the cost of developing the tool.

Business Impact

The original definition of success for this project was to increase the productivity of the IPT materials analysts (our experts). When we first met them, the analysts were spending six to eight weeks analyzing a single product. As a team, they were able to make decisions about only two or three products each quarter. They needed easier access to the data, and they needed to be able to make timely, consistent decisions from a corporate-wide perspective.

CAN BUILD enables the analyst to do a product analysis in less than a day, with far greater accuracy and detail. Decisions can be made about a specific product at regular weekly meetings or as the need arises.

The success of CAN BUILD is evident. However, the payback is only one measure of the system's impact on the corporation. The tool also enables the analyst to look at the entire product pipeline, using data that previously were not readily available and information based on calculations which were difficult or impossible to make. The result has been a more thorough picture of the potential impact of each decision. However, not even these capabilities truly measure the success of the project.

The ultimate contribution of this system is the insight it has given to the future of materials management. Once the analysts began to experiment with this simulation tool, they were able to better understand the causes of our end-of-life problem at Digital. If they could monitor and increase the synchronization of inventory throughout the manufacturing life cycle, then the end-of-life decision would be simplified. By using this same tool to periodically monitor all major products, they could reduce the amount of inventory accumulated, and thus, they could simplify, if not eliminate, the need for the write-off decisions that drove the original model.

In March 1988, six months after the initial implementation, the original vision was refocused to become a proactive "inventory goodness barometer" rather than a reactive decision support system for inventory trade-offs. Use of the tool began shifting away from responding to symptoms and toward curing the disease of excess inventory.

Until now, the analysts in Digital's manufacturing plants, with the support of IPT, had all been sharing the single corporate implementation, but we have gradually made CAN BUILD available in the plants. With eight sites running the system, we began accumulating a great deal of new expertise. We started to enhance the functionality and work toward a new, improved model. IPT satisfied its mission and began to phase out its role.

The system has proven itself, and the analysts are ready to trust it. They no longer feel the need to use the tool primarily for interactive

trial and error; they are beginning to relinquish their control and let the system do the analysis. They have begun to see the advantages of having the tool provide better recommendation and explanation capabilities and additional functions that depend on the AI knowledge base. They are now looking for more automatic features.

Conclusions

What distinguishes this AI effort from some of its predecessors at Digital (Reitman 1983) is that with this project, the developers continually compromised on the functionality they wanted to provide in favor of the functionality the users required for their business. Because we saw ourselves not only as technologists but also as agents of change, we studied and applied many change management concepts in our project decisions.

More important, we focused on the state of the need rather than on the state of the art. We allowed the reasoning component of the system to become a minor part of the full business solution and took ownership of the total solution rather than just the AI part.

Although we always tried to open the users' minds to new opportunities by prototyping more creative solutions, we listened to their requests—and often yielded to a simpler, more traditional approach. The users had little awareness of where AI was; they didn't care. We let go whenever we saw signs that the users weren't ready to take the next step. However, we left a trail of seeds behind us, and these seeds did not lie dormant for long. We focused on the business "hot buttons" that would generate the largest payback in the shortest possible time, and we used our newest technologies only where they were needed. We valued ourselves not just as knowledge engineers but also as creative software application developers, with a full workbench of tools, conventional as well as AI, at our disposal.

Postponing the development of our system's expertise was a conscious, if difficult choice that in retrospect has paid off handsomely. We see the proof of our strategy when we look at how time and experience have watered the seeds we planted in our prototyped garden; now, one year after the base-level implementation, we are beginning to see the little sophisticated flowers blooming in our user's requests. (Better yet, many of the system's users believe the ideas are their own.)

The difference was timing because now the users are ready (Conner 1985). They feel true ownership of the system and the plans for its future. We gave them a system that could be compared to a new college hire, as opposed to a Sloan fellow, and over time, they took ownership of the continued development of the new employee's expertise and awareness. Our system seems to have avoided the resistance and resentment often shown toward senior newcomers in an organization. It hasn't suffered from the classic perception of AI that it is just another textbook genius that doesn't fully understand the real world. Instead, it is welcomed as a naive but important resource with infinite potential, which the users, as the experts, are eager to train.

Each new release of the software promotes AESG's state-of-the-need system to a higher level of awareness and responsibility. Someday, as new hires are trained as materials analysts, this new generation of users might think of our system as their expert.

Acknowledgments

The author would like to acknowledge the invaluable contributions of these people to the success of this project: Ed Goucher, the IPT manager who risked investing in our project to achieve his business goals;

Mark Savage and Lucy Morini, our IPT experts, who kept faith with our process as we wrestled to define the business needs; Mike Nelson and Jim Gonyeo, the development team members whose experience, hard work, and commitment to the dream enabled us to achieve our one-year goal; John Rasku and Mike Brennan, whose fresh ideas are now carrying the project into the future; Peter Ochsner, who has managed, challenged, and supported us past every obstacle; and last but not least, Toby Frost, our technical writer, who applied her skills and enthusiasm beyond the requirements of the documentation—and this paper—to contribute to the consistency and clarity of the application itself. Together, we experienced the power and synergy of an effective team.

References

Conner, D. R. 1985. *Managing Organizational Change: Dangers and Opportunities*. Atlanta: O. D. Resources.

Cooper, T. A. and Wogrin, N. 1988. *Rule-based Programming with VAX OPS5*. San Mateo, Calif.: Morgan Kaufmann.

Gill, G. 1987. Orchestrating the Dual Perspective of Change from Theory to Practice, University of New Hampshire.

Reitman, W. 1983. *Artificial Intelligence Applications for Business*. Norwood, N.J.: Ablex.

Rockhart, J. F. and Bullen, C. V. 1986. *The Rise of Managerial Computing: The Best of the Center for Information Systems Research at Sloan School of Management, MIT*. Homewood, Ill.: Dow Jones-Irwin.

Charley: An Expert System for Diagnostics of Manufacturing Equipment

Atul Bajpai and Richard Marczewski

Machines wear in the course of performing their work. As degradation continues, machine failure is inevitable. These failures occur randomly and are disruptive to factory operations. Economic pressures force factories to reduce inventories of finished goods, and production disruptions have become increasingly costly. Industry has responded by trying to prevent these unanticipated failures through preventive maintenance programs. The premise of such a program is to repair, adjust, or replace those machine components that are subject to wear. Often, these preventive repairs are based on visual observations, and these observations are inaccurate in predicting the available life of the worn components. As a result, many components are replaced before their useful life is realized. This situation is the best case in that disruptions to production are reduced because components are replaced well in advance of failure. At the other extreme, components are not deemed worn enough to warrant replacement and then fail prior to the next scheduled preventive maintenance check. This situation causes production costs to escalate because such catastrophic failures often result in damage to subsequent downstream components. Consequently, more parts need to be replaced than would be required to replace only worn components. Finally, even the most judicious visual inspections are unable to identify faulty components that might be inaccessible.

Industry has begun to use sensor-initiated predictive maintenance checks to thoroughly determine the health of machines (Thanos 1987). A feature of sensor-initiated inspection is that quantitative data are now made available.

The use of vibration data as sensory input is commonplace now. As the technology advanced during the past 10 years, portable, fast data-collection devices became readily available (Buscarello 1987). Software advances to aid in the evaluation of these data were also realized. The problem that the user still faces is in evaluating these vibration signals to begin to list suspected faults. Then the best way to confirm the belief and narrow down the fault possibilities must be determined. It is these diagnostic skills that are so difficult and time consuming to develop. Now, with the aid of an expert system called Charley, General Motors (GM) is able to spread this diagnostic expertise to many users. This system enables even novice diagnosticians to perform at the skill level of someone with many years of experience.

Description of the Application and How It Works

The system consists of three basic modules: the knowledge base (KB), the vibration signatures (VS), and the machine database (MD), as shown in figure 1.

KB constitutes the principal module of the system and contains the knowledge about vibration analysis. The VS module consists of vibration signatures taken at different points on the machine being diagnosed. It contains frequencies and related vibration amplitudes associated with these measurement points. Vibration signatures are taken with the help of a portable, hand-held accelerometer and recorder. Finally, MD contains information about each of the major components of the machine, their relationship, and the physical features of these components.

Unlike other existing vibration-based systems (Smiley 1984), Charley covers a broad range of problems that occur in manufacturing and assembly equipment. Unbalance, misalignment, mechanical looseness, structural weakness, resonance of components, eccentricity, cavitation of pumps, problems resulting from bearing wear or failure, and problems with gear trains are typical problems that the system can handle.

Another feature of Charley is that it is able to handle different types of machines. It diagnoses problems with machines such as lathes, milling machines, drilling machines, and super-finishers. Charley is equally effective in diagnosing problems with new and old machines.

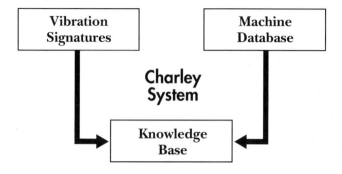

Figure 1. System Modules.

To begin a consultation, the mechanic types in "start" at the system prompt and provides the capital equipment number of the machine that is to be diagnosed. The corresponding machine description is retrieved from MD and made available to the knowledge system. The consultation continues by asking preliminary questions, such as "What are observable symptoms?" and "What does the mechanic think is at fault?" "No observable symptoms" can be answered if the mechanic wants to do predictive maintenance work, and nothing is noticeably wrong with the machine at this time. Preliminary history information is sought, and if it is available, it helps to diagnose the problem more quickly.

After the preliminary information is provided, Charley requests the mechanic supply vibration signature data for the machine being diagnosed. Figure 2 shows the signature taken on the motor of a spindle machine. With these data, Charley makes a preliminary assessment of what is wrong with the machine. Also, it determines what further tests, if any, need to be performed to confirm the faults with the machine. Once the test results are fed into Charley, the system makes the appropriate repair recommendations.

The last part of the consultation takes place after the machine is repaired. The mechanic acquires a new set of vibration signatures for evaluation by Charley to verify that the problem was indeed resolved and that the machine is running properly (Bajpai and Marczewski 1988).

At first glance, this system might look like yet another diagnostic expert system. However, several features in this application are unique. These features have made a general-purpose machine diagnostic expert system a practical reality.

Where other expert systems are application specific (Stuart and Vinson 1985), Charley is able to diagnose problems with any kind of machine. This ability was achieved with generic rules in conjunction with a machine database in which descriptions of the physical characteristics of the machine reside. MD is built on the premise that a machine is just an assembly of some common components. It recognizes that the difference between machines is simply which components are used and how they are connected. Spatial reasoning is then performed based on physical component connectivity. MD is also dynamic in that new components can be added to describe new machines. The KB architecture is such that appropriate rules are applied based on the specific machine which is presented for diagnosis. Therefore, new components can be added to MD without necessarily requiring changes to the generic KB. This kind of modularity between MD and KB and the genericness of the rules in KB make it possible to diagnose a large variety of machines with the help of a single expert system. We are not aware of any expert system that provides general coverage for machines used in manufacturing facilities.

It is important to note that significant integration issues are involved in linking KB, MD, and VS. KB consists mostly of domain rules, MD is primarily in the form of a computer-aided design (CAD) database, and the VS component is almost entirely algorithmic in nature. Various methods of system integration can be employed by program developers based on the nature of the project and the level of automation required. At the simplest level, all algorithmic calculations can be done separately, and all machine descriptions can be represented in a form suitable for direct interpretation by the knowledge engineering tool. Then, all this information can simply be transferred from external files into the cache of the tool for inferencing. In contrast, for a more automated system, proper hooks have to be built into the control of the knowledge system. It can then call the CAD and VS databases on an as-needed basis to extract the relevant data. The extracted data can be internally converted into a form suitable for inferencing. This method would be particularly useful in large operations where automation is crucial. It would avoid the users duplicating work and would help in reducing local memory requirements because much of the data are distributed across different systems.

Another feature of Charley is that any type of sensory input can be used by the system. This input includes vibration signals, forces, pressures, torques, temperatures and so on. Based on the sensory data and the information retrieved from MD, Charley is able to present the user with a hypothesized list of faults. It can operate in an automated or interactive mode and with single or multiple types of sensory data. This wide flexibility of data input and handling in both online and offline modes is not generally found in expert systems, even in those being built today.

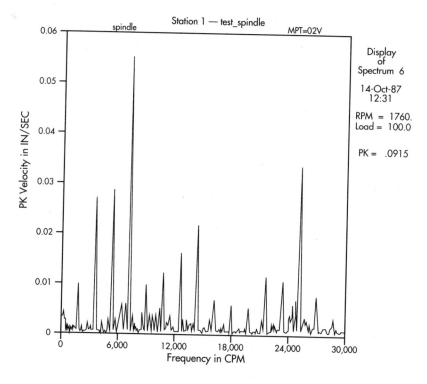

Figure 2. Spindle Signature.

If additional data are required to confirm a fault, tests are requested. The selection of the appropriate tests, as well as their ordering, is carefully prioritized. The ranking minimizes the number of tests to be performed for fault confirmation. Testing is optimized based on the number of faults that might be resolved, the degree of test difficulty, and the tools required. Additionally, tests are grouped to minimize the number of trips the mechanic is required to make to perform the tests. Related tests are grouped together. This arrangement requires the system be able to perform anticipatory thinking, in that it looks at the possible outcomes of the results and tries to optimize the efficiency of diagnosis. Most of the existing systems simply use the hypothesize-test-conclude cycle without quite looking into the overall situation.

Other diagnostic systems require baseline or trend data to determine when machines have degraded to the point of needing repair. Charley does not depend on baselines. It has predefined threshold values that are set based on the machine's precision level. These thresholds are, in fact, extrapolated baseline values, as determined by the expert. *Baseline values* correspond to values of the sensory parameters when the machine is operating in a healthy condition. Vibration signals exceeding the predefined thresholds are analyzed for fault identification. This method allows Charley to diagnose old machines even if no baselines have been established. The biggest disadvantage of using the baseline data is that in most realistic situations, such data are, in fact, never available. Therefore, systems built on this concept find limited practical use. To overcome this problem, some systems began depending on trend data. *Trend data* correspond to a series of data collected over a period of time, though none of these data necessarily have to be from the point where the machine was operating in a fully healthy condition. This method still has a problem in that the machine condition must gradually deteriorate before actual diagnosis can be performed. The method of thresholding used in Charley overcomes the shortcomings of both the baseline- and trend-based systems. In fact, with Charley, brand new machines with neither baseline nor trend information available can be evaluated and diagnosed for performance prior to being placed in service.

Charley can operate in a monitoring mode as well. Sensory data can be continually captured and evaluated. If a fault condition is detected, the system can accurately pinpoint the cause of the problem without any need for human intervention. The system can also operate in a batch mode. This ability helps avoid the large investment that is needed for automatic sensors. The key point here is flexibility; the source of the data is immaterial, be it automatic input, batch mode, or totally manual.

Unlike virtually all existing diagnostic expert systems, Charley is capable of doing predictive diagnostics. It compares current signatures with the threshold values. While performing this task, the system identifies the components that are faulty. At the same time, it identifies those components in which degradation is serious enough to warrant service while the originally identified fault is being repaired. The predictive capabilities of the system are realized by including heuristics that cause the threshold values to be suitably adjusted to find finer or impending faults.

The structuring of the system to separate the static KB from the dynamic VS and MD is also unique. The rules are developed in a generic fashion so that they pattern match with the data obtained from MD and VS. Therefore, the model used in this system can be used for virtually any kind of diagnostic application and is in no way limited to automotive manufacturing machines.

Testing, Validation, and Implementation Considerations

Once the purpose and desired functionality of Charley were formalized, several factors were evaluated to determine the plan for its development. A detailed feasibility report containing a rigorous evaluation of the suitability of AI technology was developed. It was reviewed and agreed on by the expert, the system developers, and the project managers. In testing and validating the system, one must be able and willing to assume some risks because foolproof testing is usually unrealistic. With this realization, a test plan was developed and evaluated by the users and the expert. Testing was performed on two types of cases: real cases, where the cause of the problem was already known to the mechanics and where the cause of the problem was not known to anyone, and synthesized cases, where the cause is always known because it is induced by the users.

Preference was given to testing the system on real cases. However, in many instances, the problems to be covered could not easily be located on real cases. In such situations, synthesized cases were used.

If the problem was already known to the mechanics, the system diagnosis could be verified as correct or incorrect. The expert would review all incorrect diagnoses and either suggest revisions to the rules if an error in logic existed, or give explicit reasons for the error that were beyond the scope of the system. Reasons for such errors include incorrect sensor readings; incorrect machine database entries; and at times, incorrect answers provided by system users.

If the cause of the problem was not already known, and machine repairs could not be performed to verify correctness, then the mechanic's decision on the accuracy of the system's diagnoses was considered final. Again, all incorrect diagnoses were reviewed by the expert.

Besides the field testing of the system on real and synthesized cases, regression testing was performed on an ongoing basis. This testing ensured that the changes and updates to the system did not introduce any new problems in running the previously successful cases. In addition, stress testing was performed on the system to assure that it gracefully degraded in awkward situations. Several permutations of the alternative knowledge paths were also tried. Software checks showed that over 95 percent of the code had been successfully exercised at one time or another during the entire testing period of the system.

To facilitate successful implementation of the system, an extensive training program was developed. Instead of directly training all the target end users, a few people were selected for in-depth training. Selecting the right trainers was critical to the success of the system. The selected group of trainers was representative of the typical users, and received in-depth training. In turn, these individuals taught the system to the rest of the users at various plants. Thus, most of the users learned about the system from their peers. Additionally, the trainers helped to ensure the user interface, documentation and terminology used in the expert system were in line with those of the user community.

An extremely simple and easy-to-use user interface is a requirement for any expert system. It is especially important for a system such as Charley because it is intended for use by people unfamiliar with computers or programming (Bajpai and Sanders 1987).In addition to maintaining the technical quality and functionality of the system, special care was taken to ensure that the delivery system was geared toward the target user—mechanics who perform repairs on machinery. It was made clear from the beginning that Charley was only a tool to assist, not replace, the users in doing their tasks.

Successfully Deployed System

One of the biggest disappointments in expert system technology has been the difficulty of building systems that indeed are usable. The criteria for the success or failure of the systems is simple—does it work, is it in regular use, and does it solve problems more efficiently and economically than was previously possible.

The system described in this paper was successfully deployed and is in routine production use at GM. It has been installed in many of the company's plants in distant geographic allocations. The system is sufficiently stable to be used on all shifts without any difficulty. Moreover, Charley has been in use since 1988, and to date, no significant problems have been reported. The mechanics for whom the system is intended like to use it and are easily able to cross-train each other. As a result, the user base is continually expanding. Also, the number of system installations is steadily rising. We are now planning to add new functions to significantly expand the system coverage.

Benefits

The system has the potential for tremendous savings in virtually all manufacturing and assembly plants. It can help significantly reduce machine repair costs by helping the mechanics do the following:

- To precisely identify parts of the machine that need repair
- To perform repairs or adjustments prior to catastrophic failure
- To perform repairs on a sensor-initiated basis rather than a scheduled basis

- To speed the diagnostic process
- To distribute the diagnostic expertise to many users
- To avoid fixing nonproblems.

Other benefits result from the following considerations:

- Knowing the general health of a machine helps to raise confidence in running machines unattended, alleviating fears of unexpected breakdowns.
- Charley can serve as a resident expert, on hand, 24 hours a day, seven days a week, tracking machine conditions.
- The use of the system can result in improved part quality and reduced scrap.

Because Charley applies to new machines as much as old machines, it can be a useful evaluation tool during the installation and acceptance of new or rebuilt machinery. The system can be used as a valuable training aid for persons not already proficient in machine diagnostics. It provides on-the-job training at the same time the person is working on the plant floor doing the job.

Acknowledgments

The project was initiated by the Advanced Engineering Staff and Saginaw Division of General Motors Corporation. Knowledge engineering services were contracted with Teknowledge Inc., a GM-affiliated company. System maintenance and support was contracted with Electronic Data Systems, a GM subsidiary. Help from Dr. Sam Uthurusamy of General Motors Research Laboratories in preparing this article is gratefully acknowledged.

References

Bajpai, A. 1988. An Expert System Model for General Purpose Diagnostics of Manufacturing Equipment. In *Manufacturing Review* 1(3): 180–187.

Bajpai, A., and Marczewski, R. W. 1988. An Expert System for Machine Tool Diagnostics. In Proceedings of the Second Engineering Society of Detroit Conference on Expert Systems. Detroit, Mich.: Engineering Society of Detroit.

Bajpai, A., and Sanders, B. A. 1987. Artificial Intelligence for Automobile Manufacturing. In *Encyclopedia of Microcomputers* 325–341. New York: Marcel Dekker, Inc.

Buscarello, R. T. 1987. *Practical Solutions to Machinery and Maintenance Vibration Problems.* Denver, Colo.: Update International, Inc..

Smiley, R. G. 1984. Expert System for Machinery Fault Diagnosis. In *Sound and Vibration* 18(9): 26–27.

Stuart, J. D.; and Vinson, J. W. 1985. TurboMac: An Expert System to Aid in the Diagnosis of Causes of Vibration-Producing Problems in Large Turbo Machinery. In Proceedings of the ASME International Conference on Computers in Engineering. Boston, Mass.: ASME.

Thanos, S. N. 1987. No Disassembly Required. In *Mechanical Engineering.* (September): 86-90.

Harnessing Detailed Assembly Process Knowledge with CASE

William J. McClay and John A. Thompson

The connector assembly specifications expert system, (CASE) (pronounced Casey) is an AI system that has been in use at The Boeing Company in a production mode since October 1986. It advises engineering, manufacturing, and field-service personnel in the proper assembly of electric connectors and other electric terminations that require special tools.

Detailed assembly instructions, including graphics, are printed at the user's terminal at the end of a typical consultation; these instructions provide the information needed to build a particular device for a particular program (for example, a 747 passenger jet, a B-52 bomber, or a Minuteman missile). This information is in strict accordance with the Boeing process specifications, which are contained in the Boeing Corporate Standards, various program standards, and a large number of military standards. The combined volumes of these standards contain over 100,000 pages.

Although the pages of specification data used for electric assembly probably number no more than 20,000, the process of finding the necessary information is far from simple. The specification search time for a person has been measured at an average of 42 minutes.

The system of documents to be searched can be likened to the building codes that apply to a contractor. Federal laws, state laws, county laws, and city laws all have to be considered before the contractor starts construction. Certain state laws can take precedence over federal or local laws, and various aspects of construction are found in different books. However, all laws must be considered together, applying those which take precedence. From these laws, a set of guidelines must be assembled to guide construction.

The user of the standards documents faces essentially the same, complex situation. First, all the pertinent bits and pieces of information must be collected from all the various documents and sections that might apply; then the instructions and options which do apply must be sorted out and resolved. Finally, of all the tools and materials that can be used, the researcher must decide which tools are available to the assembler and which methods will be most cost effective for the shop.

It requires years of experience to be familiar enough with the standards system to know where to look for all the relevant information and not be misled by what often appear to be conflicting statements in different sections. The ability to resolve such conflicts and provide clear and concise instructions to the shop is a rare and highly valued skill possessed by a small number of shop experts.

A previous system that attempted to automate this task collected all the bits and pieces that might apply but could not eliminate irrelevant information. It produced reports as long as 50 or 60 pages for a single connector. Another system managed to reduce the amount of irrelevant information, but the knowledge about the organization of the specifications, or specs, was distributed throughout the Fortran code that accessed the database. This complex distribution of knowledge presented a considerable maintenance problem, because specifications change frequently and, sometimes, radically.

In late 1984, when this problem was once again brought up by Boeing factory management, our Manufacturing Research and Development Organization suggested that AI might be able to contribute to a better solution. It was clear that if AI was going to provide any help for this problem, it would be in representing the standards knowledge so it was easy to capture, maintain, and reason about.

System Design

Any particular connector assembly standard is intended to answer only a small number of basic questions; for example:

- What types of connectors are covered in this document?
- What is the value of attribute X for connector Y?

- What contact part numbers can be used in connector X?
- What wire sizes can be used with contact X?
- What tools, materials, and assembly steps are required to assemble wire size X into contact Y?
- What tools and procedures are required for inserting contact X into the connector shell?
- What sealing is required for unused connector cavities?
- Are there any special problems in assembling this connector?

It seemed natural that these questions should serve as the interface between the basic consultation control mechanisms and the standards knowledge base. This setup would eliminate the need for the program that uses the knowledge to know anything about internal table structures and precedence relations between specifications, or about how to interpret any lower-level data items, which were the failings of the earlier database-oriented attempts. These interface questions would also provide a basic framework of goals that would take certain input values and, through some reasoning process, provide direct answers without any ifs in them.

This type of organization seemed to suggest an intelligent database that could ask questions when necessary to eliminate the ifs and return a set of values for each solution found. Forward-chaining systems did not seem to fit, because the user must have all solutions to a given question returned on demand. Exhaustive backward chaining seemed to be more appropriate and did not seem frightening in terms of search time, because the standards themselves limit the depth and breadth of search by explicitly referencing supplemental documents when needed. Certainty factors were not needed because process specifications are definite about what is and what is not allowed.

System Implementation

The first prototype system was built using a personal computer (PC)-based expert system building tool and proved to be satisfactory with respect to the intelligent database idea. However, a number of problems surfaced during this exercise. First, the processing time during a consultation was close to 10 minutes, which was twice as long as expected. Second, the system was required to retain certain pieces of information between consultations. Although a solution existed for this problem, using this tool was anything but straightforward .

Third, it seemed desirable to have a separate knowledge base for each standards document, so that each one could be maintained independently of the others; however, this approach was not supported. Fi-

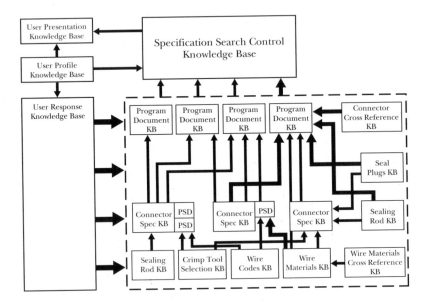

Figure 1. CASE Knowledge Base Network

nally, it was inevitable that this knowledge-based system, if successful, would be widely used throughout the corporation and would have to run on some sort of mainframe computer system. It was evident that progressing beyond this early prototype would be painfully slow unless a better tool could be found.

An evaluation was made of virtually all the expert system shells and languages then available that supported the declarative approach. Maximum flexibility, computational efficiency, and transportability were essential requirements. Because Prolog was similar to the syntax of the PC-based tool, it was tried as the new implementation language.

The effort to convert the prototype to Prolog took only a couple of days, and the results were impressive. The performance was two orders of magnitude better, and the system was now running on a VAX. No problem existed with breaking the knowledge base into separate chunks for each standard and consulting them when needed. Keeping certain consultation parameters around for more than one consultation was also not a problem.

On the down side, it took a couple of months to replace the user interface and consultation control features provided by the expert system tool. However, the rework provided an opportunity to customize the system in ways that could be extended as needed. User interface utilities were developed to provide such features as auto completion for user input and automatic menu generation from a list of items. Although still lacking in sophistication, the new shell was functionally adequate to continue the project.

The next big challenge was only beginning to be fully appreciated. The problem was that each knowledge base had to be an intelligent agent, able to interpret information being returned by other knowledge bases and selectively accept, reject, or replace any piece of information in the answer being returned. This ability required some way to label each piece of the answer at the proper granularity so that it could be easily reasoned about at some higher level.

The use of Prolog structures helped this process of filtering knowledge. For example, each text note can be represented by a note structure, such as note('BAC 5162-9', 'Strip insulation 11/16 inch and double back conductor on itself before crimping', [strip, double_back], bac5162_9fig8). This note structure identifies the text of the note as coming from document BAC 5162-9 and labels the contents as discussing both the stripping and doubling back of the conductor before crimping. The note also references a figure 8 that illustrates the operation.

These structures are easily recognized and manipulated by the pattern-matching capabilities within Prolog, which facilitated the writing of rules to deal with them. For instance, if in a higher precedence specification, using a filler wire is specified instead of doubling the conductor back, the note covering the doubling back is easily identified in the instruction list and replaced with the filler wire note along with another note to cover the stripping requirements.

Compared to a typical database system, the freedom and expressiveness of a symbolic language such as Prolog was a welcome relief and most likely a critical factor in the success of the project. Connectors vary considerably, and representing the knowledge about how they are assembled requires a flexible and expressive knowledge-modeling language. Extensive use was made of the built-in operations on lists and the powerful pattern-matching capabilities of special data structures such as the note structures just mentioned.

As illustrated in figure 1, the system architecture features a knowledge base network concept. The bold boxes and arrows represent the passing of information from knowledge base to knowledge base until the answer to the query is finally assembled and returned to the search control knowledge base. Once all the possible assembly methods are collected, search control selects the optimal choice or choices using various parameters contained in the user's profile; a report is printed on the user's local printer.

System Validation

An early prototype was installed in the shop in late 1985 and was well received; users felt it was responsive, friendly, and easy to learn. Over the next year, a considerable amount of knowledge was captured in knowledge bases, and enthusiasm for the system's capabilities and potential cost savings was widespread.

Despite this success, other obstacles had to be overcome before this technology could be widely deployed. When it was decided to conduct a formal acceptance test of Case in October 1986, a quality assurance test team selected 10 test cases and proceeded to do the research for each. No advance notice was given to the Case developers about these tests.

In six of the test cases, both Case and the test team came up with the same results. In two cases, Case contained typographic errors that were obvious to test-team members and would not have misled users. In the last two cases, the test team missed some obscure process specification departure (PSD) and indicated a tool and procedure that was not actually the latest information according to the standards.

The test results satisfied manufacturing that Case could be trusted for building hardware, and its regular use began. However, Case was not considered an engineering authority, and the quality assurance department would still have to use the paper system for its final check that everything was done correctly. However, a Case report was probably more reliable than a specification search by an experienced researcher.

Knowledge Acquisition

In order for Case to have engineering authority, it would be necessary for those engineering organizations currently responsible for the process specifications to actually produce and maintain the knowledge bases. No amount of testing would ever ensure that the hand-coded Case knowledge bases were absolutely correct. This certification problem gave rise to a knowledge-acquisition project started in 1987 that would not only take care of the engineering authority issue but would provide a significant productivity improvement for those engineers engaged in creating and maintaining standards documents.

The solution was a workstation that would include a model of a connector assembly specification and would engage the engineer in a dialog to fill in the blanks (figure 2). A Symbolics workstation running Knowledge Craft and Lisp was used for this prototype knowledge-acquisition system. After a considerable development effort in 1988, a more advanced prototype was demonstrated that solicits the necessary information from the engineer by asking for various tables to be filled in,

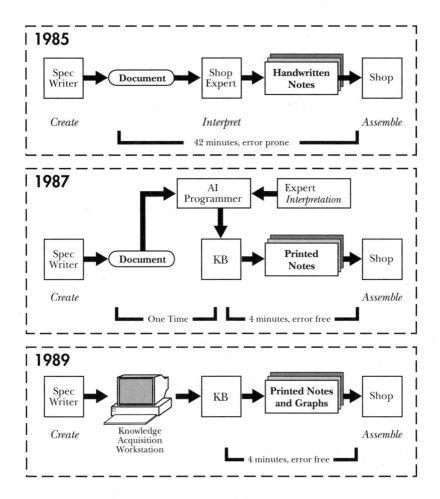

Figure 2. Automation of Standards Usage.

graphics to be scanned in, and assembly instructions to be entered into various forms that capture their underlying meaning.

The information collected by the knowledge-acquisition system is stored as a network of related objects. From this set of objects, which is called the *neutral format,* the system can produce a CASE knowledge base in Prolog or a human-readable document ready for publication in the paper system (figure 3). To achieve the document generation capability, Concordia, a Symbolics product that provides desktop publishing functionality, was integrated with the Knowledge Craft knowledge representation capabilities. Without this upcoming knowledge-acquisition

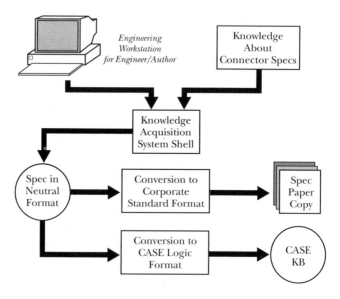

Figure 3. Standards Knowledge Acquisition System.

workstation, CASE staff members would forever be in the business of maintaining standards knowledge bases, and CASE would forever be restricted to use only as a shop aid.

Knowledge Accessibility

The most recent development project is to provide CASE knowledge directly to other software systems. Because the CASE knowledge bases can be queried much like a database, a special interface was constructed late in 1988 that provided assembly data to an IMS database application program running on an IBM mainframe. Although only a proof of concept exists, work is proceeding on a full-scale knowledge gateway with a general purpose query language that can be used for knowledge base queries as well as database queries (figure 4).

A system querying a knowledge base must be able to respond to questions of clarification from the knowledge base being queried. The reason for these questions of clarification is simply to satisfy the knowledge base's need for data as it seeks to find a solution for the stated goal. If the calling application does not happen to know the value of the attribute being asked for, it can ask the knowledge base how to ask the user for the information. In this case, the knowledge base returns

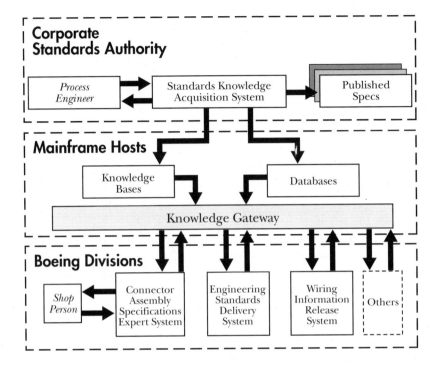

Figure 4. Future Standards System Concept.

the correct prompt string, the type of question (yes or no, single value, multiple value, and so on), a list of possible values, and a help message describing why the question is being asked and the consequences of choosing a particular answer.

It is hoped that this approach to putting knowledge on line will help solve the expert knowledge access problem and even allow applications to reason across several expert knowledge sources.

Conclusions

CASE is currently running on a MicroVAX II; it is written entirely in Quintus Prolog and has more than 25,000 rules contained in over 300 knowledge base files. It has its own expert system shell that maintains the context of the consultation and deduces answers to questions issued by the knowledge bases from previously entered information or, whenever possible, the user's profile. The shell is now being used for other expert system projects, and its capability is constantly being expanded.

CASE represents more than six person-years of effort over a four-year period. It reduces the specification search time to as little as two or three minutes and reduces the report size to as little as one page. However, most important, CASE greatly reduces the dependence on shop experts and provides the highest-quality information.

References

Brodie, M. L. and Mylopoulos, J. 1986. *On Knowledge Base Management Systems.* Berlin: Springer-Verlag.

Chandrasekaran, B. 1988. Generic Tasks in Knowledge-Based Reasoning: High-Level Building Blocks for Expert System Design. *IEEE Expert.*

Kerschberg, L., ed. 1986. Expert Database Systems. In Proceedings of the First International Conference on Expert Database Systems.

McClay, W. J., and MacVicar-Whelan, P. J. 1986. AI Based Connector Assembly. Applications of AI III. New York: SPIE.

McClay, W. J., and MacVicar-Whelan, P. J. 1988. A Knowledge Base Network for Connector Assembly Specifications. In Proceedings of the IEEE Computer Society Conference on Robotics and Automation. Los Alamitos, Calif.: IEEE Computer Society.

Marcus, S., ed. 1988. *Automating Knowledge Acquisition for Expert Systems.* Boston: Kluwer Academic Publishers.

Ullman, J. D. 1988. *Principles of Database and Knowledge-Base Systems.* New York: Computer Science Press.

Knowledge-Based Statistical Process Control

Kenneth R. Anderson, David E. Coleman, C. Ray Hill,
Andrew P. Jaworski, Patrick L. Love, Douglas A. Spindler,
and Marwan Simaan

American manufacturing has recently begun emphasizing the use of statistical process control methods to limit process variability and produce higher-quality products (Wadsworth, Stephens, and Godfrey 1986). This technology relies on the construction of charts, the observation by a machine operator or engineer that these charts indicate an out-of-control condition, a diagnosis of the cause of this condition, and the choice and implementation of a corrective action. We refer to this collection of tasks as special cause management (SCM).

Statistical process control (SPC) methods are intended to distinguish a variation in a process signal that is significantly different than the usual process variability. The SPC model assigns such variation to special causes, events that occur in time which are not part of the normal process operation. Such events might include material changes, equipment failures, operator error, or environmental changes. The unexceptional or normal variation is said to be the result of common causes of variation.

Distinguishing the special-cause variation from the common-cause variation is a probabilistic decision requiring knowledge of the normal process variability when it is in control, that is, when no special-cause variation is present. The usual technique is to apply various decision rules to control charts (Shewhart 1926; Shewhart 1931; Maragah and Woodall 1988).

Two practical problems occur that limit the effectiveness of the usual SPC approach. First, in a complex process such as aluminum sheet rolling, several process variables need to be examined. The need to monitor equipment performance, operator procedures, environmental factors, and product properties can produce an overwhelming amount of data. It is difficult for a small group of engineers to regularly examine all the data. Second, process expertise sufficient to diagnose problems causing out-of-control conditions varies dramatically from plant to plant and from engineer to engineer. Those skilled engineers who have enough process knowledge to do this work are in great demand. They often find it time consuming and tedious to manually search out causes for problems and record them for correction and historical analysis.

These limitations suggest that a scheme which combines sound statistical principles with knowledge of the process and software tools for managing a database of past performance might allow a comprehensive analysis and management of process consistency. A program called the process signal interpreters assistant (Prosaic) was developed to explore these possibilities in the manufacturing process for aluminum sheeting.

The program has two major components. The first element is a database and interactive graphics system that gives the user basic tools to examine process data. The second part of the program, which is the subject of this article, specifically addresses the process consistency problem.

Figure 1 is a block diagram of the program. A method that recognizes significant variations in process signals was developed. This method detects signal features that are impulses, mean shifts, or trends. Thresholds used in this detection scheme are analogous to limits in the usual control chart schemes. For each of the various types of out-of-control phenomena detected, a set of rules exists that is used to try to diagnose what caused this occurrence. The rules are based on past experience and basic knowledge of the relationships between special causes of variation and the symptoms they produce in signals observed in the process.

Applying the rules for each set of events results in some events being diagnosed. Information about the diagnosed events and details of their diagnosis are stored in an object-oriented database from which various summary reports can be generated. The remaining undiagnosed events can be analyzed in two ways. First, an exhaustive application of the rule set can be used to inform the user why each of the tested special causes was not asserted as the cause of the problem. This method

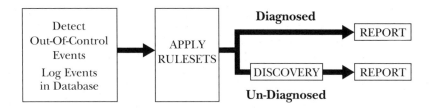

Figure 1. Special-Cause Management Block Diagram.

allows the detection of *near misses,* cases in which an event nearly meets all the tests of a given special cause apart from some exception. Second, an engineer using the program can call on an expert's view of the process, which is stored in a network diagram of influences. A method exists for searching signals in this diagram to identify other signals that were out of control when the problem occurred. The engineer can use the process model to examine the path along which the associated changes might be influencing the signal of interest.

Tools that operate on the database support the management of special causes. The engineer can examine a Pareto chart of causes for a particular signal and observe the impact of corrective actions. The engineer can also note when new special causes appear that remain undiagnosed and has tools to focus an investigation on what might cause them. Finally, the engineer can automatically generate reports to be sent to other staff members so that they are informed about the current process state of control.

Detecting Significant Variations in Time Series

Complications arise in monitoring critical process variables that constrain the validity of many control chart schemes. It is usually assumed that successive samples of the process are independently and identically distributed. Although a case might be made for this assumption in the manufacture of discrete parts, where charts have been used extensively, in many manufacturing processes , physical, chemical, and other effects introduce autocorrelation. *Autocorrelation* degrades the hypothesis testing prescribed in control charts by changing (sometimes severely) the rates of type I errors (false alarms) and type II errors (events that did not trigger alarms) (Maragah and Woodall 1988). Furthermore, the increasing sophistication of measurement and data-acquisition systems has led to higher sampling rates, which increases autocorrelation.

Additional complications arise as a result of the high cost of a broad class of corrective actions. This means that the process is not often stopped until regularly scheduled maintenance takes place. An analysis of out-of-control conditions during the run can and should be used as planning input to the maintenance session. Under this practice, the standard assumption of identically distributed data is violated in cases where out-of-control features, which change the process mean, occur in the data interval. Unfortunately, conventional control chart detections of out-of-control features after a change in the mean are unreliable and often misleading until the chart is reset. Frequent resetting is impractical.

Manufacturing process signals contain impulsive changes resulting from process shocks, sudden shifts in the mean, gradual trends, exponential decays to equilibrium and other causes. Control chart supplementary run rules attempt to detect some of these variations, but (as stated earlier) they are compromised by autocorrelation and mean changes (especially when they are superposed). In our work, we developed a nonlinear signal-processing scheme (Love and Simaan 1988) that detects (in the presence of noise) the following three basic features in a process signal: peaks, denoted by P; steps, denoted by S; and ramps, denoted by R.

Peaks are impulses of short duration, *steps* are shifts in the mean value, and *ramps* are linear trends in the data. It is assumed that useful information in the process signal can be summarized by this set of three signal features. Note that this is equivalent to modeling the data as a piecewise linear function in time, with added noise from a contaminated distribution.

The essence of this assumption is that these features represent different manifestations of process variations and, consequently, are sufficient to approximate most process behavior. Should a process variation be observed that cannot be described by some combination of these features, then the vocabulary would have to be extended and other feature detectors would be needed. An example would be a sinusoidal variation.

The automatic interpretation of signals by detecting and analyzing signal features was reported in several application areas (Stockman, Kanal, and Kyle 1976). This approach to interpretation segments the signal into regions (features) that share common statistics. Syntactic analysis techniques are then applied that treat these features like words in a sentence. More complex signal structures are generated by combining features (Fu 1974).

A block diagram of the signal-processing scheme is shown in Figure 2. The input $y(x)$ is the signal to be processed, and the outputs $P(x)$, $S(x)$, and $R(x)$ indicate the three features of interest at sample loca-

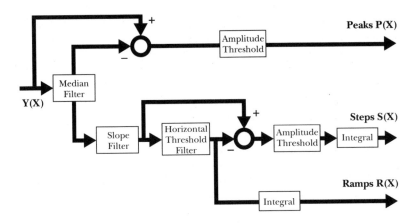

Figure 2. Nonlinear Filtering Scheme to Produce
Peaks, Steps, and Ramps from Input Signal y(x).

tion x. The basic elements of the scheme are a median filter, a slope filter, and a horizontal threshold filter.

The *median filter* is an effective technique for suppressing impulses from a signal (Tukey 1974). It replaces a sample value at location x_i by the median of sample values in a window centered at x_i. The *slope filter* replaces the sample value at location x_i with the slope of the data in the same window. This slope is determined by a linear least-squares fit to the signal values in the window. In both cases, the window is moved over the entire length of the data signal. Finally, the *horizontal threshold filter* is a nonlinear filter that replaces every sample value in the signal by the average value of all samples in a string of numbers provided: The length of the string is larger than a horizontal threshold, and the average value of the samples in the string is larger than an amplitude threshold.

To set thresholds for this scheme, the user is required to pick data intervals that are in control. These intervals are assumed to be devoid of special-cause variation, to collectively contain many points (say, 1000+), and to be indicative of the variability of the in-control process resulting from common causes. These intervals are passed through the signal-processing scheme with all thresholds set to zero. Histograms of the processed data are generated at the point where thresholding is to be applied. These histograms are assumed to represent the likelihood of observing certain values for peak heights, step sizes, slope lengths, and so on, in the in-control data; they define values for rare events.

Given a user-specified confidence limit, thresholds can be determined by calculating the area under these histograms. This area is the approximate type I error rate. The program can also estimate the type II error associated with these levels for various process shifts.

After features are detected by the nonlinear filters, a list of features is generated and parsed to determine if some subset of features should be combined as a description of a complex event. The word parsed describes the procedure of analyzing features and forming events using certain grammar rules that describe the structure of events of particular types. As an example, define a thermal, denoted by T, as a sharp step increase followed by a ramp with a negative slope (the word thermal is chosen because such an event is sometimes associated with thermal disturbances in a process). After the list of features is scanned, all features that do not participate in complex events are promoted to the same status as events. The final set of events, formed by the union of parsed events with promoted features, becomes the set of detected events for the data interval. Events are represented in the computer as objects in an object-oriented programming system called Flavors (Weinreb and Moon 1981). When events are detected, they are stored in an object-oriented database.

The operations described here can be performed interactively using a specially designed user interface. These same operations can also be run automatically by a process that starts at user-specified times and generates reports of its activities in both electronic and hard-copy form.

Diagnosing Causes for Out-of-Control Events

Rule-based diagnostic systems are commonplace in the literature of applied computer science (Hayes-Roth, Waterman, and Lenat 1983; Buchanan and Shortlife 1983). Rule-based systems are typically constructed by specifying situation-action, or if-then, statements. They possess a working memory that tracks all proven facts and an inference system which tries to prove new assertions using logical tests on facts already proven. It is assumed that the data to be tested and the knowledge used to construct rules are static and reliable. Such systems work best if the number of possible solutions or diagnoses is small. Rule-based systems are easily extended by adding rules, an important feature because new special causes often arise.

To limit the number of diagnoses and increase the modularity of the system, the overall rule base was partitioned by defining rule sets for each event type. Each rule set is used to diagnose the special cause of variation that led to this event. Such modularity makes it easier to add knowledge when it becomes available.

The diagnosis is performed by applying the appropriate rule set to a given event in a backward-chaining fashion. That is, a list of hypothesized special causes is sequentially tested by the rules. Each rule in a rule set is composed of a set of predicates that examine other data in the database to see if conditions correspond to a set of symptoms which in the past have been indicative of a particular special cause of variation.

An example rule, written in English, is as follows:

IF	there was a jam or sample taken on this lot
AND	there are product variations in the body of the lot
AND	one or more of those product variations is roughly at a distance in the lot that corresponds to the distance between machine A and machine B
THEN	diagnose this peak event as a cold spot on the material due to the material lying in contact with machine A after it was backed out of machine B.

In this rule, a *jam* is a special situation in which control is lost in the lot being processed, and the machine must be stopped. A sample is sometimes taken from a lot by stopping the machine and cutting the metal. Each of these pieces of information about the processing history of a particular lot is stored in the database. This rule expresses some process knowledge related to the distance between successive machines and how this knowledge might be related to product variations. If this set of data conditions is obtained, then the rule concludes the out-of-control peak in the signal was the result of a particular special cause—a cause that is actually an operator error in handling the lot during a jam or sample operation.

Multiple rules in the rule set come into play in proving some of the conditions. For example, a separate rule does some data analysis to conclude that product variations occurred in the body of the lot as opposed to the leading edge or the last few feet. When a diagnostic operation is performed on a set of events, the events are partitioned into diagnosed and undiagnosed sets. A report lists which causes could be diagnosed for each event. Multiple diagnoses are possible for one event.

Undiagnosed Events: Using Process Knowledge to Discover Associated Variation

New special causes will exist that cannot be diagnosed with the current rules. The undiagnosed events that result must be analyzed to discover causes and accompanying symptoms. A human expert would carry out such an investigation in several ways. A first step might be to examine other data collected, based on some model of what variations might have influenced the signal of interest. Out-of-control variations that occur at the same time as the signal variation of interest should be strong clues, which might lead to a theory of what happened. These associations are typically augmented with information that might be unrecorded (operator observations, and so on).

We developed a discover feature that uses a qualitative model of the process to look for associated symptoms in other signals. Figure 3 shows a greatly simplified association diagram, or network, that expresses the associations among the physically related variables, which lead to variation in a process parameter called bandwidth.

The internal representation of our association diagrams (also called an ANET) is a network of objects that are built using the Flavors programming system. Two basic types of objects exist: derived quantities and measured quantities. Each *derived object* can contain links to measurements and other derived objects. *Measurement objects* contain a reference to a generator procedure that can be invoked to construct a set of signal features over a given interval of investigation. In figure 3, the measured quantities are shown by oval symbols, and the derived quantities are the rectangular boxes. The diagram suggests that workroll diameter variations directly influence bandwidth and are measured by both harmonic power and roll force noise.

The links between boxes can express more quantitative relationships. For example, one signal can influence another as a simple proportion. It could also have an integral relation, where a spike in one signal leads to a step in another. Operators that express these relationships are carried on the link data structures in the network.

An *ANET* is a static, declarative representation of process knowledge. To be useful, it must be combined with a search procedure. For the purpose of discovering associated variations, a backward search through the network is appropriate. A backward search starts with the node representing the signal under investigation, that is, the signal for which an undiagnosed event was detected. A list of signal features is collected for the interval of investigation for all measurements of the starting node. These lists of signal features are then propagated recursively back through the network. When an association link is traversed, any operators attached to the link are used to transform the feature lists. When a node is reached that has measurements, feature lists for the measurements are generated and compared with the propagating lists. If a coincidence is found, an association is established and saved. The search is exhaustive, and all associations that can be reached from the starting node are considered. The result of the search is a list of signals with one or more features that are coincident with some feature of the signal

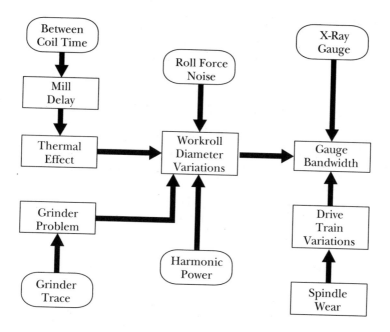

Figure 3. An Association Diagram for Bandwidth.

under investigation. Note that this result is close to the one desired when people study draftsman plots (a *draftsman plot* is a large, graphic covariance matrix where each entry in the matrix is a scatter plot of the row versus the column variable). However, cause-and-effect conclusions can generally be drawn more confidently from sets of signals that have been identified by an ANET than from draftsman plots. This confidence exists for two reasons. First, draftsman plots are limited by graphic device resolution to approximately twenty variables. Second, the ANET-based results contain consistently applied domain-specific knowledge, whereas draftsman plots are ignorant of the domain.

The associated variations of an event can be viewed as a *signature* that characterizes the event. This signature can be used to investigate the nature of the event. The most obvious use of an event signature is to focus attention on the set of signals it reveals. It is often possible to gain insight into an event by scrutinizing simultaneous graphs of the signal set in the event signature. A more intriguing possibility is the use of event signatures to classify (cluster) events. Classifying events in this manner can be a manual or an automated procedure.

Development History

In early 1985 Patrick Love was involved in a new project at Alcoa Technical Center (ATC) that required the examination of large volumes of data from an instrumented rolling machine. The objective of this project was increased understanding of the rolling process and the phenomena that affect product quality and consistency. It was discovered that a major impediment to this project was access to, and manipulation of, the online data. The data were not organized in any rational manner, and no software tools existed for the effective interactive presentation of the data. It was decided that the importance of the project warranted the development of a software tool designed to support highly interactive access and manipulation of large volumes of data.

Prosaic was developed on the Symbolics Lisp workstation. The choice of this platform was based on (1) the superiority of the software development environment, (2) the need for a high level of hardware resources to support the proposed performance of the tool, (3) the intent to add automated reasoning capabilities. An initial prototype was built within about three months. This progress resulted from enlisting the consulting services of BBN Laboratories, using a commercial software shell (KEE™), and recycling the code from an existing application.

Initially, the system was installed and used at ATC. Development and use proceeded in parallel. The initial users of the system were ATC engineers working on the rolling process consistency project. This initial use period resulted in many enhancements to the user interface predicated on user needs. In October 1986, Prosaic was installed at one of Alcoa's facilities.

The shift from a research environment to a plant environment triggered many changes to Prosaic. User support became a significant overhead issue. Previously ignored issues of interface robustness became critical. The initial implementation of the data system was found to be cumbersome, and a complete reimplementation was necessary. As part of reimplementing the data system, the frame-based data dictionary was ported from KEE to an existing, in-house frame system. Considerable effort was expended on rationalizing and generalizing the user interface. Much of the impetus for this work came from the system users. Their contributions included not only reports of system deficiencies but also suggestions for new tools and features that would aid them in their work.

Early in the development of Prosaic, it was realized that automated mechanisms for data exploration were highly desirable. The initial work on signal feature detection began shortly after the initial installation of Prosaic at ATC. This early work resulted in some general concepts around what we now call SCM (Love and Simaan 1988). The press of events delayed the extension of these ideas until September 1987 when the work described in this paper began.

The initial focus of the SCM work was to represent the physical causes of process variation as a diagram of relationships between signals. It was quickly apparent that this approach works well as a mechanism for discovering new causes of process variation but that the traditional, rule-based approach is appropriate for identifying well-understood special causes. The rule-based system for diagnosing special causes was implemented with an existing, in-house rule system together with an interface to the object-oriented representation of events. The combination of these two approaches to special-cause diagnosis and discovery, together with event archiving and statistical refinements to signal feature detection, constitute the framework for SCM.

Current Status and Future Directions

Prosaic has been in use in a plant setting for well over two years. During this time, the system has proved to be useful for many purposes, some of which are listed below:

- Investigation of out-of-control product parameters, leading to procedural changes and a reduction of the variation of these parameters (knowledge gained in this study has been implemented in SCM rules)
- A better understanding of the relationship between product metallurgical properties and process parameters
- A better understanding of limitations intrinsic to the current process equipment (this information is being used in the plan for future modernization)
- Monitoring and evaluation during the installation of new processes and comparison of trial runs to historical data
- Process and equipment troubleshooting
- Investigation of customer complaints

These accomplishments were achieved using the manual data-exploration features of Prosaic. Much of the knowledge gained during these activities was coded in the newer SCM features of the system. Much work remains before the SCM features of Prosaic begin to perform at a level sufficient to replace manual data-exploration activities. Only a small part of the process is currently embodied in the SCM tools. The system does a reasonable job of identifying special causes, although some of the detected events still go undiagnosed.

The Prosaic system continues to be actively developed. The system was recently ported to the MacIvory™ microcomputer with the objective of providing a low-cost delivery platform that can be spread to multiple sites. The data system is being enhanced and generalized to enable the application of Prosaic to domains other than the aluminum rolling process. Finally, refinement and extension of the SCM tools discussed here are a major priority for future work.

Acknowledgments

The authors wish to acknowledge the significant contribution of Mark Pate to the development of Prosaic. Mark, who is an engineer at one of Alcoa's plants, has been the major user of the system since its installation. He provided invaluable input as a computer-literate and enthusiastic user and gave the system developers numerous ideas for system enhancements.

References

Anderson, K. R. 1982. Syntactic Analysis Of Seismic Waveforms Using Augmented Transition Network Grammars. *Geoexploration* 20: 161-182.

Birman, K. P. 1982. Rule-Based Learning for More Accurate ECG Analysis. *IEEE Transactions of Pattern Analysis and Machine Intelligence.* PAMI-4: 369-379.

Buchanan, B. G., and Shortliffe, E. H. 1983. *Rule-Based Expert Systems: The Mycin Experiments of the Heuristic Programming Project.* Reading, Mass.: Addison-Wesley.

Fu, K. S. 1974. *Syntactical Methods in Pattern Recognition.* New York: Academic Press.

Hayes-Roth, F., Waterman, D. A., and Lenat, D. B. 1983. *Building Expert Systems.* Reading, Mass.: Addison-Wesley.

Love, P. L., and Simaan, M. 1988. Automatic Recognition of Primitive Signals in Manufacturing Process Signals. *Pattern Recognition* 21(4).

Maragah, H. D., and Woodall, W. H. 1988. The Effect of Autocorrelation on the Shewhart Quality Control Chart with Supplementary Run Rules. Paper presented at the Joint Statistical Meetings, ASA.

Shewhart, W. A. 1926. Quality Control Charts. *Bell System Technical Journal:* 593-603.

Shewhart, W. A. 1931. *Economic Control of Quality and Manufactured Products.* New York: Van Nostrand Rheinhold.

Stockman, G.; Kanal, L. N.; and Kyle, M. G. 1976. Structural Recognition of Cartoid Pulse Waves Using A General Waveform Parsing System. *Communications of the ACM* 19:.690-695.

Tukey, J. W. 1974. Non-linear (nonsuperposable) Method for Smoothing Data. In Proceedings of the 1974 EASCON Conference.

Wadsworth, H. M.; Stephens, K. S.; and Godfrey, A. B. 1986. *Modern Methods for Quality Control and Improvement.* New York: Wiley.

Weinreb, D., and Moon, D. 1981. *The Lisp Machine Manual.* Cambridge, Mass.: MIT Press.

Logistics Management System: Implementing the Technology of Logistics with Knowledge-Based Expert Systems

Gerald Sullivan, Kenneth Fordyce, and LMS Team

The problem of scheduling a semiconductor fabrication facility (fab) is difficult for at least two reasons. First is the high degree of inherent combinatorial complexity. Given the number of processes and products, pieces of manufacturing equipment, and personnel in a typical fab, a large number of individual lots assigned to specific machines at particular instants in time are possible. Second is the high degree of uncertainty associated with the execution environment. In the longer term, fabs ramp processes and products up and down, replace old machinery with new, and reassign experienced personnel. In the shorter term, the delicate nature of the process and equipment and changes in product orders generate frequent scheduling surprises (Kempf, Chee, and Scott 1988).

Within the complexity of semiconductor manufacturing, three related decision areas can be distinguished based on the time scale of the decision window. The first decision area, FabStrategic scheduling, contains a set of problems that stretch weeks and months into the future. Here, decisions are made about the impact of changes in the product distributions, the amount and type of equipment in the fab, personnel assignments, and ramping up of new processes. The second area, FabTactical scheduling, deals with problems of the next shift or two, or of the next few days. Here, decisions are made concerning trade-offs between running engineering and manufacturing lots, prioritizing late lots, positioning preventive maintenance downtime, and grouping products to reduce setup time. The third area, FabReal-Time scheduling, addresses the problems of the next few hours. Decisions here are made about the inevitable surprises that occur when the schedule produced in the fab tactical stage is executed (Kempf, Chee, and Scott 1988; Kempf 1989).

We refer to tactical and real-time schedule decisions as manufacturing dispatch." Making good dispatch decisions in large manufacturing operations requires the successful use of application information systems to implement the manufacturing logistics that improve manufacturing performance.

Overview

The logistics management system (LMS) is a real-time, imbedded, transaction-based, integrated decision and knowledge-based expert support system that serves as a dispatcher for the manufacturing flow of IBM's semiconductor facility in Essex Junction, Vermont. This plant produces a variety of logic and memory micro-electronic chips and modules. The purpose of LMS is to help improve manufacturing performance in throughput (tool utilization and cycle time) and serviceability (delivery of product on schedule) by supporting tactical and short-term decision making. Significant success was achieved in both areas. LMS is now a critical component in running major areas of the manufacturing facility.

LMS regulates the complex manufacturing tasks of an entire plant. It automatically picks up data, and using the knowledge given it by scores of human experts, it reasons so thoroughly about manufacturing production and makes corrections and changes based on this reasoning so quickly that no individual or group of individuals can match its performance (Feigenbaum, McCorduck, and Nii 1988, p.64).

The Complexity of the Logistics in Making Chips

Consider the problem: An operator on the fabrication line is responsible for the work-in-process stream through a number of fabrication tools. Some of the tools—the photolithography machine, for example—are expensive, perhaps costing a million dollars or more; so there are relatively few. Other tools are cheap and plentiful. Chip wafers flow quickly through the low-utilization, cheap tools and then get in line to go through the highly utilized expensive tools, the so-called pinch-point, or bottleneck, tools. This lineup delay causes imbalances in the production line, which can actually start the work in process oscillating because as the wafers wait to go through the expensive pinch-point tool, the photosensitive coating begins to deteriorate.

The operator is responsible for all this and other worries too. The tools the operator is responsible for can't all run the same kind of products, and because the operator has a mixed stream of products, thought must be given to which product is going where. Furthermore, tools are idiosyncratic about when and how they age and what effects this process produces and so on. Meanwhile, the operator has to remember *process specifications*, the sequence of fabrication steps. Each wafer must go through, say, a photo process 10 or 15 times. Each process builds a layer resulting in a three-dimensional complexity—you can't build level B before you build level A. This complexity is compounded by other difficulties; for example, boxes, as the wafers are called, look alike, and if an important customer comes in with a special urgent request, then everything is up for grabs. For operator and manager alike, the whole manufacturing process is like a giant chess game, except the complicated sequencing, combinatorial complexity, and the unpredictability make the manufacturing process harder to think through than even the most demanding of games (Feigenbaum, McCorduck, and Nii 1988).

The LMS Base

The pre-LMS decision support environment and the LMS base consisted of several items. First was a set of automated but independent data systems, which reliably (strong data-integrity checks, wanding, hardware backup) recorded all transactions in real-time to lots, machines, and orders in manufacturing; provided basic process flow checks and information about the manufacturing process; and established (overnight) the priority of a lot once a day. These systems are critical to the manufacture of chips. The facility has zero tolerance for their malfunction.

Second was a set of paper systems that contained process specifications, machine specifications, lot locations, operator ability, and so on. When they were developed, these systems were state of the art. However, they were independent (limited coordination between them) because of historical development, technical ownership and maintenance, and technical barriers to linking them. From a decision support point of view, the databases should not be independent but logically linked.

Some of the limitations of the existing systems included limited reporting capability; no real-time access to the data to support management, industrial engineers, or operators; limited operator assistance on assigning lots to tools; and inability to adapt schedule to changes in operation real time.

Implementation of LMS

As the size, activity level, and product mix in the facility grew, the limited decision support provided by these systems was no longer sufficient to meet the needs of the business. The Essex Junction facility remained data rich but was becoming information poor. The advanced industrial engineering (AIE) department was missioned to produce LMS as a first step to reverse this trend.

The first requirement (step 1) was to gain real-time access to the different systems; generate a logical integration of the several pieces of data; use a knowledge base to enhance the information content of the data, where appropriate; and provide a single location to obtain real-time access to the integrated data. This requirement drove the architecture to be transaction based. The current transaction rate is 240,000 per day. The LMS component that carries out this task is called the Gateway. Gateway keeps a record of the current status of each lot and each machine.

The next requirement (step 2) was to provide various users or customers (managers, operators, planners, and AIE) with tools that quickly, flexibly, and in real-time converted the data to information in the form of a "view." In this step, paper knowledge bases were computerized and integrated into the data stream. This portion of LMS is called the management access technique (MAT). MAT put the users into the decision support phase.

People quickly recognized that the time (decision) window, in which an opportunity could be leveraged or a problem avoided, was small; the transaction rate generated cognitive overload on the human expert; and the existing experts had no well-articulated underlying theory to improve logistics performance. Therefore, more responsibility had to be moved to LMS and the AIE department, and the knowledge for LMS would come from a team of people with various specialties.

The next step (step 3) was to establish proactive intervention with alerts. Knowledge about what situations should be cause for alerts and who should be alerted was encoded in a KBES, and these modules monitored the data stream. The alerts were made operational through MAT. This knowledge did not come from a single expert but from re-

CHIPTYPE	STAGE	→ SETUP
tiger	1	3
tiger	2	2
lion	1	4
lion	2	4

Table 1 (T1).

SETUP	→ SETUP_TIME
2	20
3	20
4	50

Table 2 (T2).

CHIPTYPE	SETUP_TIME	→ PROCESS_TIME
tiger	2	60
tiger	3	50
tiger	4	NA
lion	2	NA
lion	3	NA
lion	4	60

Table 3 (T3).

search into the manufacturing logistics with a team from the industrial engineering, production, and manufacturing engineering areas, and management. This LMS component is called the Alert.

Step 4 was to extend proactive intervention to the action phase by having LMS respond to anomalies and make the dispatch decision about what lot to run next. The knowledge for this activity had the same source as step 3. This LMS component is called the Scheduler. Although LMS is fully implemented, work is continuing to improve the function provided at each step.

In a real sense, LMS is imbedded in the stream of transactions coming from a variety of online, real-time control systems, developing and

```
∇PM1 [ □ ]∇
       ∇
 [0]   PM1
 [1]   ⱷ THIS FUNCTION DETERMINES
 [2]   ⱷ THE APPROPRIATE SETUP CONDITION
 [3]   ⱷ
 [4]   ⱷ VALUE CALCULATED (OUTPUT VARIABLE):      SETUP_COND
 [5]   ⱷ
 [6]   ⱷ POSSIBLE OPTIONS FOR OUTPUT VARIABLE:    LONG
 [7]   ⱷ                                          SHORT
 [8]   ⱷ
 [9]   ⱷ VALUES USED ( INPUT VARIABLE):      SETUP_TIME
[10]   ⱷ                                     PROCESS_TIME
[11]   SETUP_COND← ⊂ 'LONG'
[12]   CONDS←(SETUP_TIME<25),(SETUP_TIME<4 x PROCESS_TIME)
[13]   → (∧/CONDS) / L010
[14]   → 0
[15]   L010:
[16]   SETUP_COND ← ⊂ 'SHORT'
[17]   →0
```

Figure 1. Procedure Module 1 (PM1).

implementing recovery tactics without human involvement in the decision process. Its role is not cognitive replacement but cognitive augmentation. Not a single expert can do what LMS does. Its underlying theme is "bring the appropriate knowledge at the appropriate time to the appropriate place to capitalize on an opportunity before it disappears."

LMS Goals and Functions

The first goal of LMS is to reduce the time to locate lots. On the manufacturing floor, many lots are stored in identical looking boxes at each operation. Being able quickly find a specified lot is a critical first step. LMS provides decision support tools to assign storage locations for each lot (trying to put a lot near the machine likely to process it) and display the storage location of a specific lot on demand.

Goal 2 is to ensure express lots are processed quickly. A lot is designated express because of a pressing order from a manufacturing location or a test run sent by the Essex Junction development laboratory. LMS provides decision support tools that track and report on the progress of these lots and attempt to minimize their waiting time.

The third goal is serviceability (delivering the product on schedule). LMS provides decision support tools that manage lots which are plus (ahead of schedule) or minus (behind schedule) to schedule.

Goal 4 is to improve tool utilization and reduce cycle time. LMS provides decision support tools that manage retooling time, set up *trains* (lots requiring identical machine setup for bottleneck or pinch-point machines), appropriately mix lots with different test requirements, maintain appropriate buffers before and after pinch-point tools, control preventive maintenance and prioritize repair actions, and match the lot with the tool best able to handle the lot.

Goal 5 is to ensure unnecessary WIP levels are not generated. LMS provides decision support tools that control the launching of lots into the next sector, remove lots, and watch buffer levels.

The sixth goal is the most difficult: coordination between goals. LMS provides functions that attempt to assess the trade-offs between competing goals. For example, an express lot arrives at a bottleneck machine that is not set up to run this lot but is set up to run a set already waiting.

Structure of the LMS Scheduler

The LMS scheduler is the portion of LMS that watches the transaction stream and recommends an action to take. This component of LMS is structured as follows:

Each of the goals of LMS (processing express lots, controlling WIP, enriching the service rate, and enhancing tool availability) is broken down into a set of transaction data requirements, calculations on these data, a set of fact bases or tables, and a set of heuristic function modules. This combination is called a goal advocate.

Each goal advocate monitors the flow of data and uses its components to turn the data into information. When a machine becomes available, it uses this information to develop an opinion about what action to take next to best achieve its goal. The opinion includes not just its wish about what to do next but the severity of the need and the impact of alternative choices. This opinion or view is communicated to the judge. The judge listens to the opinions and makes a decision about what to do next that is in the best global interest of the facility.

For example, assume the manufacturing process has four steps or operations (S1, S2, S3, and S4). Lot E122 is an express lot that is arriving at operation S2. The express advocate (EA) knows from its knowledge base that the preferred tool is T2-10. EA then checks the availability of this tool and for other express lots that might be waiting for this tool. If this tool is available now or will be soon, and no other express lots are waiting for this tool, then EA recommends assigning Lot E122 to tool T2-10.

Now assume EA determines two other express lots (E234 and E245) are waiting for tool T2-10. First, it calculates the slack available at this step based on due dates and estimated times to complete steps S3 and

S4 for each lot. Second, from its knowledge base, it determines if tool T2-10 is the preferred tool for these lots. If EA finds lot E122 has less slack than lots E234 and E245, and tool T2-10 is not the preferred tool for E234 and E245, then EA gives lots E122 priority over lots E234 and E245 for tool T2-10.

As a second example, assume the manufacturing process has four steps or operations (S1, S2, S3, and S4). Operation S3 is the bottleneck operation. Three machines are at S3: T3-10, T3-20, and T3-30. The tool utilization advocate (TUA) might recommend running lots L1227, L1228, L3447, and L3420 as a group through tool T3-10 to optimize the use of this tool. The serviceability advocate (SA) might insist lot L6776 must run today on tool T3-10 to avoid critically falling behind schedule. The Judge can explore a number of options; one that often occurs is asking TUA to find some lots with the same setup requirement as lot L6776. This balances the need to meet the schedule for lot L6776 and avoiding a setup on tool T3-10 for a single lot.

Knowledge Representation and Use in LMS

The representation, storage, and manipulation of knowledge (see Fordyce, Norden, and Sullivan 1989 for a definition of knowledge and knowledge-based systems) is an important part of each major component of LMS. Examples of knowledge use in the Gateway include characterizing a transaction (is it a lot movement, a change in status of a machine, a change in an order), converting a field stored in HEX to a character, and calculating elapsed time. Elapsed time would be defined as the calculation of the current time minus the elapsed time with adjustments. Adjustments can be made for machine availability, second shift work, a holiday, and so on. In MAT, knowledge about the relationships between operations is used to structure a view, and knowledge about the relationships between data fields is used to provide assistance in making queries. In the Alert, knowledge is used to specify a situation that requires the attention of a decision maker. For example, if the machine xx is idle for more than 30 minutes during regular shift hours, then alert the floor supervisor. Examples of knowledge use in the LMS Scheduler include the setup required for a lot at a specific iteration, machine preferences, the lower and upper bounds on the time needed to complete a wafer, the identification of lots requiring the same setup, and the choice between conflicting goals.

In LMS, a variety of techniques are used to represent, store, and manipulate knowledge. In each case, the underlying theme is a function-object representation of knowledge (FORK) scheme. This approach

CHIPTYPE	STAGE	→ SETUP_TIME
tiger	1	20
tiger	2	20
lion	1	50
lion	2	50

Table 4 Composite 1 (TC1).

	T1	T2	T3	P1
CHIPTYPE	1	0	1	0
STAGE	1	0	0	0
SETUP	0	1	1	0
SETUP_TIME	0	0	0	1
PROCESS_TIME	0	0	0	1
SETUP_COND	0	0	0	0

Table 5 INMATIP.

uses the object-oriented programming concepts of avoiding data-type dependencies, linking data and procedure to generate a natural object, and being robust to change. Additionally, FORK draws on concepts from functional programming languages (such as APL2 and Lisp) and mathematical function notation. A detailed description of FORK can be found in Fordyce and Sullivan (1988b). It appears Rish et al. (1988) and Goldbogen (1988) have independently developed approaches similar to FORK.

Tables and Program Modules as Functions

Tables, or fact bases, represent a tabular representation of a functional relation between input and output variables, where the domains and ranges are a finite set of elements. Program modules (PMs) are small programming modules used to describe functional relationships that carry out standard conditional logic and computation on the input variables to generate the output variables. The links between functions represent composite function operations. An example set of functions is shown in tables 1-3.

Table 1 is the mapping between the input variables CHIPTYPE and STAGE into the output variable SETUP. Table 2 is the mapping between

	T1	T2	T3	P1
CHIPTYPE	0	0	0	0
STAGE	0	0	0	0
SETUP	1	0	0	0
SETUP_TIME	0	1	0	0
PROCESS_TIME	0	0	1	0
SETUP_COND	0	0	0	1

Table 6 INMATOP.

Variable	Value
CHIPTYPE	tiger
STAGE	2
SETUP	unknown
SETUP_TIME	unknown
PROCESS_TIME	unknown
SETUP_COND	unknown

Table 7 Variable Values.

the input variable SETUP into the output variable SETUP_TIME. Table 3 is the mapping between the input variables CHIPTYPE and SETUP_TIME into the output variable PROCESS_TIME. Figure 1 is the mapping between the input variables SETUP_TIME and PROCESS_TIME into the output variable SETUP_COND.

These relationships can be written in the following functional notation:

SETUP = T1 (CHIPTYPE, STAGE)

SETUP_TIME = T2 (SETUP)

PROCESS_TIME = T3 (CHIPTYPE,SETUP_TIME)

SETUP_COND = PM1 (SETUP_TIME,PROCESS_TIME)

The concept of a composite function does not exist with our table and procedure module method of describing functions. For example, the functional relationship between the input variables CHIPTYPE and STAGE and the output variable SETUP_TIME can be found using table 1 and table 2 and viewing the variable SETUP as an intermediate output variable:

SETUP_TIME = TC1 (CHIPTYPE,STAGE) = T1 ∘ T2 .

LOT_FAMILY	→ PRIORITY	→ MASK
tiger	5	brown
lion	10	blue

Table 8 (T4).

OPR_FAMILY	→ MACH_FAMILY
bend	xxx
bake	yyy
test	zzz

Table 9 (T5).

MACH_FAMILY	→ SETUP_TIME
xxx	2
yyy	1
zzz	3

Table 10 (T6).

The concept of algebraic simplification can sometimes be done by generating a new table. The table composite (TC) function TC1 would result in table 4.

In APL2, tables map directly into two-dimensional general arrays. Table 4 can be generated with the following statement:

T1← 4 3 p 'tiger' 1 3 'tiger' 2 2 'lion' 1 4 'lion' 2 4 .

T1 is a matrix with a shape (p) of four rows and three columns. The matrix is filled in row by row (row major).

APL2 provides indexing into any portion of the matrix and a variety of comparison operations. The following statements access column 1 and check if any element in column 1 is equal to lion:

$CoL1←$ T1 [;1] .

The variable col1 is assigned the values in column 1 of T1. This statement stores the values in column 1 of T1 in the variable col1. Col1 is a vector with four elements.

The statement

$MATCH1←$ (⊂'lion') ≡ $COL1$

matches (≡) the character string lion against each element in the variable col1. match1 is a vector with four elements, one for each member of col1. An element of match1 gets a 1 if the corresponding element of col1 has lion, else a 0. match1 is 0 0 1 1.

In APL2, PM can be executed at any time with the execute primitive (⍎ or ⎕ EA).

The Network of Functions

Using the following two Boolean arrays and some Boolean array primitives in APL2, we can determine how the different functions relate to one another (Fordyce and Sullivan 1987a; Fordyce and Sullivan 1988b; Jantzen 1989).

The first item generated is a Boolean matrix called INMATIP (IP is for input). This matrix records which variables are input variables for which functions. INMATIP has one row for each variable and one column for each function (table or procedure module). A cell gets a 1 if the variable is in the input portion of a function, else a 0. For this example, INMATIP would be as shown in table 5.

The second item generated is a Boolean matrix called INMATOP _ (OP is for output). This matrix records which variables are output variables for which functions. INMATOP has one row for each variable and one column for each function (table or procedure module). A cell gets a 1 if the variable is in the output portion of a function, else a 0. For this example, INMATOP would be as depicted in Table 6:

With these two arrays, we can determine that the variable SETUP_TIME is a function of the variables CHIPTYPE and STAGE through the intermediate variable setup. The following APL2 expression gives us this information:

W ← INMATIP ∨.∧ ⍉ INMATOP

W ← W ∨.∧ W .

Using a slightly more complicated set of Boolean operations, we can generate an ordering of the functions based on relative independence. To explain relative independence, let's look at an an example; if you had the equations

Volume = area x ht	(1)
Perimeter = (2 x length) + (2 x width)	(2)
Area = length x width	(3)
Heating cost = 4 x volume ,	(4)

you would need to execute equation 1 before equation 4 and equation 3 before equation 1.

We could view equations 2 and 3 as making up the most independent group or class of rules because their input variables (length and width) are not calculated by any other equation. Equation 1 would be in the second group or class because its input variables are not calculated by another equation (ht) or an equation already ordered (area). Equation 4 would make up the third group.

For our example, the functions are ordered as follows:

Class 1: T1
Class 2: T2 T3
Class 3: PM1 .

This kind of information is particularly useful in debugging, firing more than one function per inference cycle, and resolving conflicts.

An Overview of Inferencing

Assume we initially know the information displayed in table 7.

From the information in table 7, we generate a Boolean vector KUVEC. This vector tells us whether a variable has a value. One cell exists for each variable. A cell gets a 1 if the variable has a value, else a 0. For this example, KUVEC is 1 1 0 0 0 0.

Given these Boolean arrays, a fast and efficient forward or backward chainer can easily be implemented in APL2 (Fordyce and Sullivan 1988a, 1988b; Jantzen 1989).

The underlying premise is as follows: We only want to execute a function if values exist for the input variables. Using KUVEC and INMATIP, we can identify candidate functions to fire with the following APL2 expression:

Poten ← ∧ / [1] KUVEC ≥ [1] INMATIP .

Integrating FORK and Transactions

In the following example, we illustrate how FORKs are integrated with the LOT tracking database and the transaction stream:

In our manufacturing facility, we have lots that we process, operations which we do to the lots, and machines that carry out the operations. For each lot that is launched into the manufacturing stream, we keep the information shown in Figure 2.

When a lot is launched into production, the first two variables (LOT_ID and LOT_FAMILY) in the record are given values. The value for the variable OPR_FAMILY is changed in the record each time the lot enters a new manufacturing operation. All the other fields are generated from one of the function objects depicted in tables 8-11

These relationships can be written in the functional notation shown in table 12.

Procedure Module 2

output variable: EST_LV

input variable: PRIORITY

                       OPR_FAMILY

this procedure module estimates the expected time the lot will leave the operation where it is presently located.

Table 11.

PRIORITY	= T4 (LOT_FAMILY)
MASK	= T4 (LOT_FAMILY)
EST_LV	= PM2 (PRIORITY, OPR_FAMILY)
MACH_FAMILY	= T5 (OPR_FAMILY)
SETUP_TIME	= T6 (MACH_FAMILY)
SETUP_TIME	= TC2 (OPR_FAMILY) = T6 ∘ T5 (OPR_FAMILY)

Table 12.

Table composite 2 (TC2)

opr_family	-> setup_time
bend	2
bake	1
test	3

Table 13.

The composite function TC2 can be simplified into the fact base displayed in table 13.

The values for PRIORITY and MASK are obtained from T4 when the lot is launched. Because lot 11129 belongs to the FAMILY tiger, it has a priority value 5 and a MASK value brown.

The value for SETUP_TIME is obtained from T5 and T6 each time the lot enters a new operation. From T5, we obtain the machine used in the operation, then we obtain the SETUP_TIME for the machine from T6. For lot 11129 entering the bend operation, MACHINE is xxx and SETUP_TIME is 2.

The value for EST_LV is obtained from executing PM2.

The lot table now has the record for lot 11129 shown in figure 3.

LOT_ID	LOT_FAMILY	OPR_FAMILY	PRIORITY	MASK EST_LV	SETUP_TIME
11129	tiger	bend	5	brown 300	2
11130	tiger				
11132	lion				

Figure 2. Lot Tracking Database

Notes and Observations

Basic concepts and techniques from MS/OR/STAT such as statistical estimations, risk analysis, the resolution of conflicting goals, and decision trees play a major role in LMS. Especially pertinent was a view that humans had significant cognitive limitations that negatively affected their decision-making ability. Because of the complexity of the logistics problem (generated by time to produce results, data engineering, complexity of the product flow, number of products, short decision windows, and so on), traditional MS/OR/STAT structures such as mathematical programming and multiattribute decision making were insufficient for globally structuring of the knowledge and taking incremental steps in developing a solution. KBES-type heuristic model structures provided the only realistic vehicle for development of the model (see Fordyce, Norden, and Sullivan 1987a and 1987b for a discussion of KBES as a model). Real-time, imbedded decision support was the only realistic vehicle for delivery. Five points should be noted:

First, a key feature of KBES structures is the ability to do rapid prototyping and incremental evolution. These features permitted the LMS industrial engineers to add components to deal with the complex interdependencies between goals, production steps, and decisions made over time as they came to understand their nature and relative importance. These features will also permit the IEs to adapt the system in real time as underlying conditions at the facility change.

Second, much of the knowledge put into LMS was the result of STAT/OR/IE kinds of investigations by the AIE department in meetings with LMS customers (called XSELL sessions). The managers and operators often did not articulate specific rules but goals and relative trade-offs that AIE helped turn into operational guidelines. The IEs had access to the real-time flow of events on the line and could change rules within 30 minutes. Therefore, they could put in heuristics, watch their impact, do some analysis, and then repeat the process.

Third, for a system to continue to survive, it must continue to evolve. To maintain a steady pace of evolution, LMS relies on user-friendly programming tools such as rules, views, menus, and fact tables to permit the end user to carry a significant portion of the burden for writing and updating the system. The enclosure and message-passing concepts from object-oriented programming were key in permitting the system to evolve at a rate equal to customer demand.

Fourth, the data engineering and networking are as difficult as the knowledge base development and MS/OR analysis.

Fifth, as work on LMS progressed, it became apparent that general KBES tools would not suffice for this task. This type of tool puts the knowledge collected in rules and frames into a single bucket, and the inference engine determines which knowledge to use when. This type of tool does not permit explicit procedural control over the activation of knowledge (we do not consider flag variables in a rule explicit control), does not provide for the addition of different types of knowledge representation schemes, often requires a lot of central processing unit memory just for initialization, is at best difficult to imbed, and makes it almost impossible to implement give-and-take reasoning.

LMS was developed using APL2, Pascal, and Assembler to build our own knowledge-based expert system environment (Fordyce and Sullivan 1988a, 1988b). APL2 is a particularly potent programming language for building RITIDES systems. It provides for rapid prototyping, strong data array structures generalized across numeric and character data, portability between host and workstation, and full access to all elements of the computer system (Brown 1988, Brown et al. 1987; Fordyce and Sullivan 1987b, 1988c).

The LMS team comprised people with strong system thinking skills and a strong knowledge of manufacturing (either industrial engineers or people directly from manufacturing), strong programming skills, and strong skills in KBES and MS/OR/STAT.

LMS was developed over two years with a team varying between four and six people. The two years included the time to move the system from a limited implementation with a few users to a full implementation with over 400 users.

References

Arnoff, E.; Kania, E.; and Day, E. 1958. An Integrated Process Control System at the Cummins Engine Company, *Operations Research* 6(4): 467-498.

Brown, J. 1988. *APL2 at a Glance*. Englewood Cliffs, N.J.: Prentice-Hall.

Brown, J.; Eusebi, E.; Fordyce, K.; and Sullivan, G. 1987. APL and Expert Systems. *AI Expert*. 2(7): 72-86.

Chedzey, D.; Holmes, D.; and Soysal; M. 1976. System Entropy of Markov Chains. *General Systems*. 21: 73-85.

Feigenbaum, E.; McCorduck, P.; and Nii, P. 1988. *The Rise of the Expert Company: How Visionary Companies are Using Expert Systems to Make Huge Profits.* New York: Times Books,.

Fordyce, K., and Sullivan, G. 1988a. Boolean Array-Based Inference Engines. IBM 34EA/284. Kingston, N.Y.

Fordyce, K., and Sullivan, G. 1988b. Table-Based Function Mappings and Knowledge Representation. IBM 34EA/284. Kingston, N.Y.

Fordyce, K., and Sullivan, G. 1988c. Why APL for Advanced Decision Support. IBM 34EA/284. Kingston, N.Y.

Fordyce, K. and Sullivan, G. 1987a. A Boolean Array-Based Algorithm in APL for Forward Chaining in Rule Based Production Expert Systems, 185-19. In Proceedings of the APL87 Conference. 16(3). New York: Association for Computing Machinery.

Fordyce, K., and Sullivan, G. 1987b. APL and Extending Decision Support Systems with Expert Systems and other New Technologies. In Proceedings of Expert Systems in Business Conference. Medford, N.J.: Learned Information.

Fordyce, K.; Griesinger, D.; and Sullivan, G. 1985. ACDS: Adaptive Cybernetic Decision System. IBM 34EA/284, Kingston, N.Y..

Fordyce, K.; Norden, P.; and Sullivan, G. 1989. One Definition of Knowledge Based Expert Systems. IBM Tech. Rep. 34EA/284, Kingston, N.Y.

Fordyce, K., Norden, P., and Sullivan, G. 1987b. Links Between Operations Research and Expert Systems. *Interfaces.* 17(4): 34-40.

Fordyce, K.; Norden, P.; and Sullivan, G. 1987a. Review of Expert Systems for the Management Science Practitioner. *Interfaces.* 17(2): 64-77.

Goldbogen, G. 1988. Technical Note: Rules vs. Methods, Rensselaer Polytechnic Institute. Troy, N.Y.

Goldrat, T.; and Cox, J. 1986. *The Goal: A Process of Ongoing Improvement.* Croton-On-Hudson, N.Y.: North River Press.

Jantzen, J. 1989. Inference Planning Using Digraphs and Boolean Arrays. In Proceedings of the APL 89 Conference. 19(3). New York: Association for Computing Machinery.

Kempf, K.; Chee, Y.; and Scott, G. 1988. Artificial Intelligence and the Scheduling of Semiconductor Wafer Fabrication Facilities. *Sigman Newsletter.* 1: 2-3.

Kempf, K. 1989. Manufacturing Planning and Scheduling: Where We Are and Where We Need to Be. In Proceedings of the Fifth IEEE Conference on AI Applications, 13-19. Los Alamitos, Calif.: IEEE Computer Society Press.

Posner, M. 1973. *Cognition: An Introduction.* Glenview, Ill.: Scott, Foresman.

Rish, T.; Reboh R.; Hart, P.; and Duda, R. 1988. A Functional Approach to Integrating Database and Expert Systems. *Communications of the ACM* 31(12): 1424-1437.

Sullivan, G. 1976. EIN: Executive Information Network. IBM 34EA/284. Kingston, N.Y.

Wrege, C.; Greenwood, R.; and Peterson, P. 1986. Achieving Stability in Automobile Production with a Pioneer MIS: 1908-1917, 157-159. In Proceedings of the 1986 Northeast Association of Decision Sciences Annual Regional Conference. Baltimore: Johns Hopkins.

Woolsey, G. 1988a. The Fifth Column: On the Optimal Fueling of Airlines, or There's No Fuel like an Oil Fuel. *Interfaces* 18(2): 72-74.

Manufacturing Design

Coolsys: A Cooling Systems Design Assistant
Chrysler Motors Corporation and Texas A&M University

Design Advisor: A Knowledge-Based
Integrated Circuit Design Critic
NCR

OHCS: Hydraulic Circuit Design Assistant
Kayaba Industry Company, Ltd.

Coolsys: A Cooling Systems Design Assistant

Patricia G. Friel, Richard J. Mayer, Jeffery C. Lockledge,
Gary M. Smith, and Roger C. Shulze

Engineering design in the automotive industry is subject to two prominent characterizations. First, engineering design is a highly distributed
activity with complex, time-consuming mechanisms for managing the
distribution of requirements and the assimilation of design components. Second, engineering design is highly prototype driven. In most
cases, no methods exist for determining the adequacy of a design.
Hence, iterative prototype construction and evaluation is the standard
method for producing a final design. Both characteristics, although
unavoidable given the current state of the art, contribute to long design-cycle times, typically three years.

Not surprisingly, the reduction of design-cycle time is a major management objective in automotive engineering. The production of better initial designs (before the construction of physical prototypes begins) is one way to reduce time in design. Another point of possible
speedup is in the rapid reevaluation of designs in response to engineering change notices. Another significant management objective typically addresses uniformity in engineering methods because it is expected that a product with a higher, more consistent quality will result.
The application described herein addresses both of these objectives.

A Cooling System Design Assistant

Developed by the Knowledge Based Systems Laboratory at Texas A&M University for Chrysler Motors Corporation, the cooling system design assistant (Coolsys) is an integrated set of tools for engineering design support in the cooling systems domain. The system functions within the prototype-and-evaluate paradigm for engineering design, the prototypes constructed being computer models of vehicle component functionality. Coolsys incorporates three modes of operation: expert, sensitivity analysis, and manual. These modes are implemented not as separate programs but as an integrated set of tools that the engineer can pick up and put down almost at will while using the system. The expert mode of operation was the primary focus of this work; however, the other two modes were included to recognize that effective automated design support must not restrict an engineer to those tools deemed intelligent, but should provide broad support for the design process.

In the *expert mode,* the system generates design specifications for engine box cooling systems given a description of the vehicle from the cooling systems point of view, that is, a description of the related subsystems, such as engine, air conditioning system, and transmission. Test conditions (for example, speed and ambient temperature) and certain technical or administrative constraints on components are also input to the reasoner. The reasoner then generates as many adequate design solutions as it can given the heuristic capabilities it has. Design proposals are generated using a combination of general domain knowledge and of knowledge specific to the problem case at hand.

As each design proposal is generated, it is evaluated using a Fortran engineering analysis program known as the thermal and airflow model (the thermal and airflow model was written some years ago by Roger C. Shulze in the cooling systems lab at Chrysler). This program models the performance of a cooling system given a system description and a description of the other subsystems that affect the performance demands on the cooling system. The thermal and airflow model returns a data set of performance indicators that the reasoner uses in deciding what modifications to make to a proposed design. The iterative process of redesign and test is complicated by the need to find a design solution which is satisfactory under multiple test conditions which tend to work against each other; for example, added shroud increases airflow at idle (a positive effect) but might increase coolant temperatures at high speeds (a negative effect). Thus, the reasoner must track the design configurations and accompanying test results that it has previously tried in order to make tradeoffs among these conflicting goals. The system frequently finds multiple solutions and occasionally finds none.

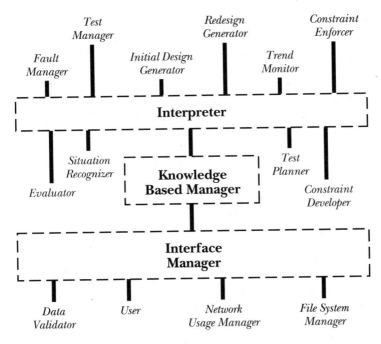

Figure 1. Generic Functional Components

In the *sensitivity analysis mode,* Coolsys gives the engineer a tool for experimentation. In this mode, the engineer can specify repeated runs of the thermal and airflow model varying some parameter (either of the cooling system or some other vehicle system) over a range. Thus, the engineer might, for instance, study the effects of different body styles on cooling system demands or verify that a design found satisfactory by the expert mode is in fact a stable design; that is small changes in a design feature do not produce large changes in performance.

In the *manual mode,* the thermal and airflow model can be run as a stand-alone program. This mode is sometimes desirable if an engineer wants to make a quick check on some proposed vehicle design change. The manual mode is also useful when an engineer wants to experiment with an unusual design feature, something outside the purview of the expert system's knowledge, but that can be simulated using the thermal and airflow model. In a future version, the manual mode might provide the base structure for a knowledge-acquisition tool that would be used to capture design rationale as new product components or technologies are incorporated into cooling system engineering practice.

From a knowledge-based system view, Coolsys addresses two basic problems. The first problem is modeling the reasoning process used by an expert design engineer as a design is iteratively proposed and evaluated against acceptance criteria. The reasoning process was effectively modeled using a situation-specification technique that operates in the context of a history of design experimentation. The second problem is integrating symbolic and numeric computing components. Major issues involved the understanding and handling by the expert system of errors arising in the analysis program and techniques for allowing the expert system to understand the assumptions underlying the analysis model.

A generic functional architecture for designer systems of this type was developed, and Coolsys was implemented according to this architecture. The basic functional components are illustrated in figure 1 and are explained in detail in Friel (1988).

Coolsys is written in a combination of three languages: the automated reasoning tool (ART™) for the reasoning model, Lisp for various procedural components, and Fortran for the engineering analysis model. The rationale behind using ART and Lisp was simply that each was used for the purpose for which it was designed; Fortran was used for the performance model because the program will be maintained by mechanical engineers versed in Fortran.

The Design-Reasoning Model

The primary innovation in the Coolsys work is the situation-based model of engineering design reasoning implemented in the Coolsys reasoning component. *Situation based reasoning* is the process of using a current event viewed in the context of an ancestral chain of past situations and against certain background conditions to determine the current situation. In the design context, the current event comprises a paired design decision and design evaluation. In an implementation of the situation-based reasoning model, each current event must be individually accessible by the reasoner, separately, but the situation representation includes the current event, a chain of past events, and some number of underlying background descriptions. Background descriptions are layered so they can be viewed as a coherent unit but still remembered separately.

The recognition of situations is basic to the way the Coolsys reasoner is structured. The importance of this system structure was not immediately obvious when we first interviewed the expert, Gary M. Smith of Chrysler Cooling Systems. We first heard comments such as "this should be done before that," implying some prioritizing scheme. However, further analysis revealed that an engineer will immediately abandon a personally stated prioritizing scheme if a situation arises that doesn't fit the general case. Apparently, the priorities do represent a compiled, or abstracted, knowledge that certain actions are generally more effective (either technically effective or cost effective) than others. An experienced design engineer will readily produce a flowchart representation of how personal decisions are made. However, our experience was that specific design problem cases never fit the abstraction exactly. It is the completely specified situations that over time have given rise to the abstraction, which must be captured to effectively model the engineer's reasoning process.

With the understanding that the objective is the capture of design situations, the problem becomes one of designing a knowledge base structure that enables this capture. Two particular characteristics of this design process affected the eventual structuring.

The first characteristic is the utilization of experimentation history. Design situations and resultant decisions can frequently only be determined in the context of a history of design experimentation. As an example (figure 2), suppose that the analysis routines of the thermal and airflow model were run on a cooling system design with a specified fin density of 20 for the radiator. The analysis predicts a coolant temperature of 259 degrees fahrenheit entering the radiator. The goal temperature is 250 degrees. Given this situation alone with no background information, the apparently correct decision would be to increase the fin density to provide increased copper surface to dissipate heat from the radiator. However, if this situation is played knowing that previously a fin density of 18 yielded a coolant temperature of 255 degrees and that the general fin density versus temperature curve is parabolic, one can conclude that fin density should be decreased to allow more air to pass through the radiator. Thus, the maintenance of the design experimentation history is necessary to intelligently assess the current situation.

The second characteristic is the one situation–many design options phenomenon. Typically, in any design situation, multiple design changes could make sense. The engineer might want to try more than one option, possibly comparing the results. Thus, it is desirable that multiple design situation histories be maintained in parallel.

To accommodate these characteristics, the knowledge base was structured into a hierarchically ordered set of situation representations. This structure allowed the one design–many design options characteristic to be modeled. It also permitted rules to be written such that pattern matches on a current situation could be evaluated in the context of previously existing situations. ART Viewpoints™ used with the ART production-system paradigm, provided the underlying language structure to implement the situation-based reasoning model.

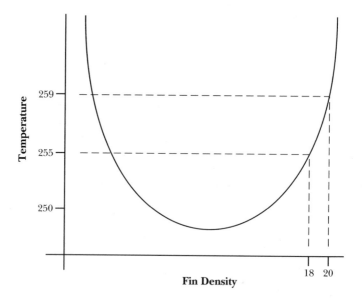

Figure 2.

Effectively, Coolsys is able to learn the special behavioral characteristics of the vehicle being designed by viewing the design development and testing history. It learns what design options advance (or do not advance) the design goals and uses this knowledge to dynamically refine constraints on design parameters. That is, the technical constraints imposed by the user (or by default) at program initiation can be modified by the program itself as it learns the behavior of a specific vehicle. Tightened constraints narrow the space of probable design solutions and hasten convergence to an acceptable solution.

This model is a distinct departure from previously proposed design models that have either viewed the design process as a quasilogical process (Hillier, Musgrove, and Sullivan 1984; March 1984) or as constraint propagation, search, hierarchical decomposition, and so on (see Brown and Chandrasekaran 1986; Mittal and Araya 1986; and Steinberg 1987). Much has been written about how to model design in general, or mechanical design in particular. Our experience in the engineering design arena, however, leads us to believe that most models that have been proposed suffer from overgenerality. The design process is probably not amenable to a single definition because many different reasoning processes take place in generating a design. The situation-based model that we implemented is only one of these. Evidently, certain forms of qualitative reasoning and curve-based reasoning (which we are investigating in the Knowledge Based Systems Laboratory) are also prominent in engineering design.

The Integration of Symbolic-Numeric Components

Our work also explored the problems inherent in a tight integration among a design reasoner, engineering analysis models, and test databases, essentially the symbolic-numeric computing problem (Kitzmiller and Kowalik 1987). Two faces of the problem were evident in the Coolsys work. The first is the practical problem of integrating languages primarily designed for either symbolic or numeric computing but not both. The difficulties are apparent in Coolsys, where the integration of ART and Lisp is natural and unobtrusive, but the incorporation of the Fortran program is far from elegant, although workable. The basic problem is that the integration of languages with the two orientations has not been recognized by the designers of expert system languages as a major issue. In engineering design, however, the incorporation of existing analysis programs, which most likely will be written in Fortran, can be expected to be the norm.

The more serious integration difficulty arises when the reasoner in the symbolic world needs to understand what is going on in the numeric routines. It is frequently insufficient for the reasoner to treat an engineering analysis program as simply a black box that returns results, because results can be incorrect or reflect incorrect assumptions for a variety of reasons. Additionally, error conditions can arise within the analytic code, and normally these conditions are not directly available to the reasoner. Certain error conditions in the thermal and airflow model, for instance, clue the knowledgeable engineer to certain input data problems, for example, too much trailer weight or an incorrect tire rolling resistance, but these associations are not obvious to the uninitiated.

The expert system, like the knowledgeable engineer, should be able to make these associations, report them to the user, and recover gracefully. An innovation in Coolsys was a context-sensitive approach to the identification of, and recovery from, such conditions. The fault monitor component in Coolsys watches for problems reported from the thermal and airflow model, and suggests possible causes of error conditions to the user. Because of the layering of background conditions and the maintenance of the design history, the system is able to return the knowledge base to a state where recovery options are possible. The structure of the knowledge base accommodates recovery well; however, the discovery of problem conditions in the analytic code still depends on reporting by the code itself. The problem of recognizing what is happening in the analytic code is inherent to the symbolic-numeric computing dichotomy.

Criteria for Success

Our criteria for a successfully deployed application include the following:

- The application should be judged cost effective by the organization. Benefits can be assessed in a number of ways, such as direct dollar savings, reduced training time, increased effectiveness of the personnel using the system, and better first-time designs.
- The application should be smoothly integrated with other systems that interface with the application, for example, information systems, database systems, or predictive analysis programs.
- The application's users should find the interface natural to work with; that is, the interface should make the system fit unobtrusively into the user's work style.
- The application should be maintainable and extensible by the user organization.
- The application should be extensible as new computer technology emerges or as new problems in the domain arise.
- Users in the domain regularly use the system as an integral part of their work process. User acceptance is often indicated by a steady stream of requests for system modifications or enhancements.

Payoff to the Organization

The benefits to the organization in the case of Coolsys are several:

- Reduction in the time to generate an initial design: Coolsys is able to generate design specifications in a few minutes as opposed to the several days it previously took.
- The rapid generation of multiple acceptable designs: Coolsys is frequently able to generate a number of satisfactory designs. This ability enables an engineer to choose a best design from several, whereas in the past, time constraints prevented the engineer from developing more than one acceptable design.
- Solidification of the engineering method: One of the objectives of the Cooling Systems Department was to better understand how it performed the design task as a group. The knowledge engineering exercise helped formalize the department's engineering methods, thus addressing the management objective of uniform engineering methods.

- Technology capture: The organization gained experience in how to select future expert systems applications and how to manage its development and deployment.

- Identification of reasoning patterns commonly used in engineering design: These concepts will expedite the identification and development of future systems in engineering.

Project History

The development of Coolsys to its initial deployment was a two-phase project, essentially, a problem selection and prototyping phase and a development phase. The first phase was a 6-month (8 person-month) project, roughly half of which was devoted to the problem-selection process. The second phase was a 12-month (18 person-month) project that focused on working with an expert designer to develop the prototype reasoning module and with potential system users to develop the user interface. The system has been in use in Chrysler Cooling Systems since 1987, and all cooling systems for new vehicles (except for trucks) are being designed with the aid of the system. Coolsys is now being maintained and expanded by Chrysler, with the engineer who served as the expert now in charge of system maintenance.

In a sense, the system is still being deployed. At the time of its initial deployment, only one Symbolics 3645 was available for use. However, four MacIvory systems are currently being purchased and will be distributed at convenient locations for access by the engineers. Also, the user interface is being revised in response to user requests. It is expected that this evolutionary process will continue as new capabilities are added to the system and new technology is absorbed.

Acknowledgments

The authors would like to acknowledge the invaluable contributions to the project by M. Sue Wells and Paul Squitterri of Texas A&M University.

References

Brown, D. C. and Chandrasekaran, B. 1986. Knowledge and Control for a Mechanical Design Expert System. *Computer* (July):92-100.

Friel, Patricia G., 1988. Modeling Design Reasoning in Automotive Engineering, Ph.D. diss., Computer Science Dept., Texas A&M University.

Hillier, B.; Musgrove, J.; and Sullivan, P. 1984. Knowledge and Design. In *Developments in Design Methodology*, ed. N. Cross, 250-258. New York: Wiley.

Kitzmiller, C. T. and Kowalik, J. S. 1987. Coupling Symbolic and Numeric Computing in Knowledge-Based Systems. *AI Magazine* 8(2): 85-90.

March, L., The Logic of Design. In: *Developments in Design Methodology*, ed. N. Cross, 265-272. New York: John Wiley.

Mittal, S., and Araya, A. 1986. A Knowledge-Based Framework for Design. In Proceedings of the 5th National Conference on Artificial Intelligence, 856-865. Menlo Park, Calif.: American Association for Artificial Intelligence.

Steinberg, L. I. 1987. Design as Refinement Plus Constraint Propagation: The VEXED Experience. In Proceedings of the 6th National Conference on Artificial Intelligence, 830-835. Menlo Park, Calif.: American Association for Artificial Intelligence.

DesignAdvisor™: A Knowledge-Based Integrated Circuit Design Critic

Robin L. Steele, Scott A. Richardson, and Michael A. Winchell

NCR Microelectronics provides the necessary software, service, and manufacturing capability to support integrated circuit (IC) designers in the creation of Application Specific Integrated Circuits (ASICs). These ASICS are often designed by engineers with varying levels of experience, from novice to expert, and varying backgrounds, from IC designer to system designer.

The knowledge required to design and implement a successful ASIC is distributed among diverse experts who perform specific tasks related to ASIC design and manufacture. This knowledge includes expertise from design engineers, product and test engineers, and quality engineers, all with specific knowledge of various aspects of IC design and manufacture. However, this expertise is often unavailable to the designer during the specification and design of a component and only becomes available at the time of a design review or during prototype testing. Changes to a design at this stage of the process are costly in terms of time and money; therefore, it is desirable to correct problems as early as possible in the design cycle.

DesignAdvisor provides a mechanism for capturing this distributed knowledge, representing years of experience, and allows the designer to apply it to a design. The system provides suggestions regarding specific aspects of IC design quality while the design is still in progress, and the cost of implementing changes is relatively low. The system analyzes partial or complete designs, pointing out critical problems that might cause the device to fail or be difficult to manufacture and test and provides helpful guidelines and information to improve performance and density.

The DesignAdvisor approach is a departure from most research in applying AI technology to problems in the VLSI design domain, which has typically focused on some form of design synthesis or decomposition (Mostow 1985). Integrated computer-aided design (CAD) systems that automate design synthesis from a high-level design description into a prefabrication database will be the elite CAD tools of the future. However, until the difficult problems facing such a system, such as low-level performance analysis, redundancy minimization, and intercommunication among synthesis tools (Gu and Smith 1986) are solved, much can be gained through intermediate knowledge-based applications such as DesignAdvisor (Steele 1987).

DesignAdvisor is in commercial use and has provided valid advice that designers have chosen to follow for each design on which it was run. Payoffs for the designer include increased productivity (a correct design much sooner); lower design costs; first-pass success (advice uncovers fatal errors); and higher-quality, more consistent designs.

Description of the DesignAdvisor System

As an ASIC vendor, NCR Microelectronics is exposed to a wide variety of designs with widely varying style, complexity, and function. With this variance comes a need for increased sophistication in analysis. DesignAdvisor analyzes performance, testability, and manufacturability of CMOS semicustom VLSI designs implemented using a library of digital and analog cells. Input to the system consists of a netlist (text) description of a circuit logic schematic. Once the circuit information is represented within the knowledge base, additional information is obtained when needed through interactive dialogue. Hierarchical design methodology, a must for managing complexity in VLSI design, is preserved by running the system on net lists for design modules as well as complete designs. DesignAdvisor runs on the design engineer's workstation (initially on the Mentor Graphics/Apollo DN3000/DN4000 series) and is integrated into the designer's accustomed environment where it highlights problem areas within the schematic capture system.

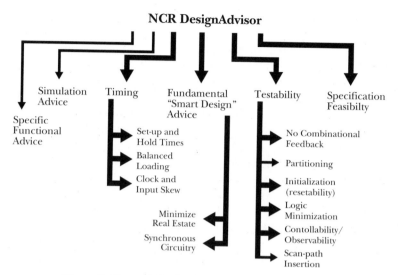

Figure 1. Example DesignAdvisor Attribute Hierarchy. Note that highlighted paths have been addressed .

An initial hierarchy of design attributes, which serves as a framework for the system's recommendations, was compiled by studying the major problems discovered in commercial designs over the past several years (figure 1). This analysis was conducted to pinpoint typical mistakes that had caused unsatisfactory silicon prototypes. It found that the attributes of DesignAdvisor address a high percentage of these problems. Sixty-six percent of the design problems in first-pass prototypes would have been detected by a system possessing the attributes of DesignAdvisor. Seventy percent of the timing-related problems alone could have been avoided. This sometimes subtle area accounted for the majority of the observed problems. The skeleton hierarchy has expanded throughout the development of the system to encompass attributes in greater detail.

The system was implemented using a prototype hybrid expert system development tool called Proteus (Petrie, Russinoff, and Steiner 1986). The tool integrates both known and novel AI programming techniques into a robust vehicle for the implementation of knowledge-based systems. Proteus integrates a logic-based rule mechanism, including both forward- and backward-chaining rules, with frame-based object representation and a fully integrated truth maintenance system (TMS) based on the work of Jon Doyle (1979). Lisp s-expressions can be used in both the antecedents (preconditions) and the consequent (resulting actions) of rules.

A TMS-Based Default Reasoning Approach to Design

Design systems typically reason with large amounts of incomplete knowledge. This form of knowledge is inherently intractable and cannot be applied within reasonable resource bounds (Levesque 1986). By representing and reasoning about partial knowledge in a principled way, it can be made tractable. Forcing a complete representation of the knowledge domain can provide a context for guiding search and limiting combinatorial explosion. Incomplete knowledge can be made more manageable by using default reasoning to fill in missing detail and closed-world assumptions to define clear boundaries for the domain.

Consider a scenario where one is seeking technical advice from a knowledgeable but possibly imperfect design consultant. The exchange would probably involve an explanation of the existing design and maybe some additional question-and-answer dialogue. On obtaining the facts and recognizing how the design related to past experience, the consultant can then offer some advice on how to improve the design. In general, this method is effective for obtaining feedback on the evaluation and refinement of a design approach. The exchange can allow for communication of concerns and viewpoints between the consultant and the designer for use in focusing the analysis. The focus can then be targeted for the designer's accustomed fashion of design rather than forced into overly restrictive design practices. The educational and productive exchange can also force consideration of critical trade-offs in the design and point out possible deficiencies in the design methodology. In addition, the advice can be overridden with knowledge contrary to that of the consultant.

This model of interaction with the NCR DesignAdvisor illustrates the use of default reasoning for an advisory approach to design (Steele 1988). Normal and plausible design practices can be addressed first, using simplifing closed-world assumptions for domain guidance, and complex situations can be handled on contradiction of default advice, using domain-guided, dependency-directed backtracking (DDB). This approach has shown both the effectiveness and the limitations of using a justification-based TMS for problems with large search spaces where several applicable solutions must be found. In contrast to an assumption-based TMS (ATMS)[1], the control provided by domain guided DDB can be used to focus the search on subsets of the solution space to avoid needless inefficiency. By using this form of control, the system can be directed toward the efficient detection of constraint violations and the subsequent identification of faulty assumptions. The strategy has proven to be effective for reasoning about the design process in general because it closely corresponds to the designer's own thought process.

A Novel Application of Contradiction Resolution

Applying the default reasoning approach, the system advises on problem areas using incomplete knowledge, making reasonable default assumptions based on typical logic design techniques. Initially, the system considers only a small number of exceptions to the default assumptions in each advice area. Less common exceptions that apply to specialized cases are considered only when a designer disagrees with the advice given. If the designer contradicts the default advice, the system can present known, less common alternatives to the designer to determine if a justification exists for retracting the advice. The integrated TMS provides the mechanism for identifying and retracting faulty assumptions (Petrie 1985). to resolve the contradiction of a default design rule violation. The contradiction resolution itself is performed using the DDB control mechanism. Frequent criticisms, involving the lack of control of this technique, were overcome by allowing the contradiction resolution to be guided by domain knowledge (Petrie 1986).

The system is prepared to provide exceptions to each advice category, although these exceptions might or might not cover the user's particular case. If an appropriate exception cannot be determined by the system, the user can still oppose the advice, but the user is prompted to explain the exception to provide a record of the decision in the session transcript. This explanation helps designers communicate the deviation from typical practice to themselves and colleagues and provides an opportunity for the knowledge engineer to continually improve the knowledge base through the use of the recorded sessions and additional consultation with expert designers. This method allows for rapid inclusion of additional exceptions to each area of advice and, therefore, minimizes the knowledge-acquisition bottleneck.

Knowledge Base Examples

The knowledge base is broken down into modules that correspond to specific areas of expertise. This knowledge was elicited from interviews with senior designers from among our experienced customers and interviews and case studies with NCR design, product, and test engineers. These experts brought their experience to bear on five broad categories: timing and clocking, speed performance, system interface input-output (I/O) specifications, general knowledge about good design practice, and testability. Output of each module is in the form of advice concerning a number of design aspects. The designer then has the option to seek further explanation about a piece of advice, reason with the system regarding possible exceptions to the advice, or simply accept the suggestion and make the recommended changes.

We believe that much of the value of the system's advice lies in its availability to the designer during actual design and schematic capture of ASIC, independent of simulation. This early availability can reduce the necessity for tedious work with conventional tools and potentially eliminate costly redesign, substantially cutting the time to market. The following examples are drawn from recent DesignAdvisor sessions on designs in progress and illustrate key system capabilities.

Timing-Clocking Expertise Module

The timing-clocking expertise module prompts the designer for the names of the system clocks and associated timing data. Derived clock signals and clock periods are then determined and automatically propagated throughout the design. The system then checks for problems such as clock skew, glitch-prone and risky gated clocks, insufficiently buffered clocks, and excessive clock speed. Risky asynchronous practices are advised against because of a preference for synchronous design. In addition, information is provided about critical paths and routing efficiency factors.

An example of timing advice provided by the system is as follows:

Phase-skew-use-PCL2
ERROR—The clock phases N-3055 and CLKX3 (to the DFFR I-802) are skewed. No more than an inverter difference is allowed. A PCL2 should be used in this situation.

In this instance, the system warns the designer that the possibility of a clock skew exists. This possibility was determined by looking at the relationship of the signals driving the clock input of the clocked logic cell and noting that more than one inverting buffer lies in the path between the clock and clock-bar signals. This heuristic allows the system to quickly identify potential clock-skew situations.

Performance Expertise Module

The performance expertise module gives advice on topics such as optimal buffer configurations (serial versus parallel), placement or removal of buffers to speed critical paths, the loading of buffers that can improve routing inefficiency, and optimization of buffer placement when complemented signals are available. High-drive cells (logic cells identical to standard logic gates but capable of driving larger capacitive load) are recommended when appropriate. These analyses eliminate tedious manual calculations and reduce the need for guesswork or repeated simulations to determine the optimum buffering of signals in the design; for example:

Replace-buf-cell

NOTE—The buffered cell NOR2H I-12 provides no speed up for the signal N-32—The buffered cell should be replaced with the non-buffered equivalent NOR2.

This example indicates a situation where the designer placed a high-drive cell (NOR2H—a two-input inverting logical OR) on a path where it was not actually needed. Circuit paths without a large amount of capacitance are actually slower if driven by a buffered cell. The system can determine the best way to correct the situation, recommending the use of a smaller-drive cell in this case. Similarly, the system can recommend use of buffered cells or buffers on paths that drive large loads, always with the goal of optimizing the area/speed trade-off. This knowledge module uses a complex tree-pruning algorithm as it seeks to meet highly interactive constraints.

System Interface Expertise Module

The system interface expertise module verifies proper I-O pad and buffer combinations to meet system requirements. When evaluating system requirements, several attributes of each signal, such as noise levels or tristate requirements, are considered. Also, the power needs of the entire design are considered to verify and, if necessary, improve the designer's choice of the number and placement of power and ground pins. This is an area where many designers make common mistakes that can go unnoticed until prototypes fail to function within the intended system.

General Expertise Module

The general expertise module identifies poor design practices such as feedback loops (using combinational logic cells) and delay logic. These practices are especially dangerous in standard cell design because with the use of automated routing, more uncertainty exists in predicting delays than in full custom layout. Other errors checked by this module include floating tristate buses, unconnected input signals, and incorrect use of pull-up and pull-down elements and other special cells. An example of advice from this module is as follows:

Floating-Bus-Error

ERROR—The output signal OUT[1] of the tristate element I-21 #1 may not drive logic unless a pullup or pulldown is used to guarantee that a floating bus will never exist.

In this example (figure 2), DesignAdvisor looks at the connections of each cell, checking for common design errors related to the intended use of this cell. This example illustrates a rule that might not be obvious to a new designer and might even be overlooked by an expert but

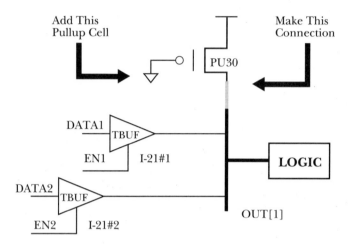

Figure 2. A Tristate Bus Must Be Pulled Up Or Down to Guarantee That the Bus Will Not Float.

is absolutely critical to the success of a CMOS design. Tristate buses allow two or more devices to drive the same signal at different times, dependent on the state of the control signals to these devices. Thus, not driving a bus signal or allowing it to float to an undetermined value causes any fault on this bus to be undetectable.

In this situation, the designer needs to add pull-up or pull-down devices to each floating bus. These sorts of problems might be considered minor in terms of how little trouble they are to correct, but they are easy to make and difficult to find if the design covers several sheets of a schematic. In the past, these errors might have been caught by peer reviews of a design or at a structured design review involving the design engineer and NCR engineers responsible for manufacturing and testing the completed IC. DesignAdvisor can serve as an automated design review, thus reducing the burden of managing the complexity of large designs.

Testability Expertise Module

The testability expertise module currently contains knowledge pertaining to the controllability of clocked logic elements. Designs are more easily testable when each clocked logic element can be conveniently set or reset. For example, given the names of all external reset signals, the system traces internal resets and then looks for deeply embedded resets, embedded nonresettable clock elements, and invalid resets that are not related to an external reset signal:

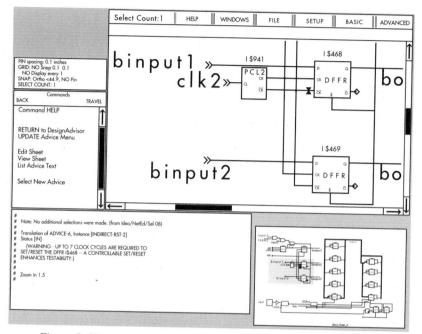

Figure 3. The DesignAdvisor System is Coupled With the Schematic Capture Evironment so that Problem Areas Can be Pointed Out Visually.

Indirect-Reset

WARNING—Up to ten clock cycles are required to set or reset the clocked logic cell I $468. A direct set or reset enhances testability.

In this case (figure 3), the worst-case path to set or reset this particular element I$468 can require 10 clock cycles, meaning that the clocked logic cell is potentially difficult to control. An overall system reset should be able to put the entire design in a known state; this ability reduces test time for an individual circuit and increases the overall testability of the circuit, thus lowering the overall cost of the part to the customer.

As an example of the default reasoning capability of the system, consider the situation where the designer wishes to contradict this advice. Although the advice, based on heuristic knowledge about controllability in typical IC designs, is valid in a local sense, the designer might have global design knowledge that renders the advice invalid. In this case, the system attempts to resolve the contradiction by seeking some new information about this exception, responding with a query:

Direct control of reset signals is preferred. An excessive number of states may be required to control the gated reset signal—Can the signal be controlled functionally with only a few input patterns?

Suppose the designer knows that the element can easily be set or reset by directly loading in a value during test mode and responds positively to the query. The system then changes the state of the database to indicate it no longer believes this particular piece of advice is true, justified by the user's explicit knowledge about the situation. If this case is not so, and the designer accepts DesignAdvisor's recommendation, then the circuit should be redesigned to more easily reset this clocked element.

The advice provided by DesignAdvisor's five expertise modules is currently of commercial quality and has been used by a number of customers. Additionally, DesignAdvisor system provides a flexible foundation for incorporating new knowledge as it is gathered from users (both internal to NCR and external to its customers) and systematic knowledge acquisition in new areas of expertise.

Development History

A large-scale prototype was designed at NCR, by the Advanced Development Group. In a joint effort, the Software Development Group subsequently generated a product from the prototype. During product generation, top-down software development methodologies were employed (Mintz 1989).

The initial prototype system was developed on a Symbolics Lisp machine and required 3.5 person-years of effort. The product development effort that followed the advanced prototype system required an additional 5 person-years of effort and was accomplished in 1 year by using the original Advanced Development engineers as well as three Software Development engineers. Software Development staff members served internships in the Advanced Development group to be trained effectively and create a cooperative team for the intense product development effort that followed.

The system has been used internally on real designs since October 1987. The system was run by NCR staff on customer designs for a period of nine months prior to an external product offering in July 1988. All new internal designs done within the Microelectronics Division now undergo a DesignAdvisor analysis.

Commercial Success and Payoff: Designer Feedback

DesignAdvisor has been used on many designs, and in each case, the designer received valuable advice that was taken into account prior to

device prototyping. In 40 percent of the commercial designs analyzed internally, *first-pass failure* (prototypes that are not fully functional in the target system as a result of designer error or oversight and require redesign) was avoided by following the advice given by DesignAdvisor. This savings is significant in both time (about 12 weeks) and cost (on the order of tens of thousands of dollars).

Other success measures for the customer include improved design quality; more consistent design practices; tutorial advice; and improved documentation of the critical, esoteric portions of the design. Other measures of success for our organization include a position of industry leadership in providing state-of-the-art technological solutions to meet customer needs and a more cost-effective environment for designing ASICs. This tool is showing financial impact well beyond the revenue generated through software sales and maintenance by enhancing NCR's leadership position in the ASIC industry.

Acknowledgments

The authors would like to thank the MCC Expert Systems Group, headed by Charles Petrie, and Dr. Dan Ellsworth (NCR) for their consultation and support. Special thanks are also owed to Eric Mintz (NCR), Chris Whitley (NCR), and Dr. Andrew Ressler (Liszt Programming Inc.).

Notes

1. Such approaches have advantages over the Doyle-style TMS for certain types of problems, but ATMS control mechanisms are weak (deKleer 1986).

References

Doyle, J. 1979. A Truth Maintenance System. *Artificial Intelligence* 12(3).

deKleer, J. 1986. Problem Solving with the ATMS. *Artificial Intelligence* 28 (2).

Gu, J., and Smith, K. F. 1986. KD2: An Intelligent Circuit Module Generator. In The Proceedings of the IEEE International Conference on Computer Design: VLSI in Computers.

Levesque, H. J. 1986. Making Believers Out of Computers. *Artificial Intelligence* 30(1).

Mintz, E. 1989. Knowledge Base Development: A Top-Down Approach to Design and Test. Paper presented at the Rocky Mountain Conference on Artificial Intelligence, 8–9 June. Denver, Colo.: RMCAI.

Mostow, J. 1985. Towards Better Models of the Design Process. *AI Magazine* 6(1): 44-57.

Petrie, C. 1985. Using Explicit Contradictions to Provide Explanations in a TMS, Microelectronics and Computer Technology Corporation Technical Report, MCC/AI/TR-0100-05.

Petrie, C.; Russinoff, D.; and Steiner, D. 1986. Proteus: A Default Reasoning Perspective. In Proceedings of the Fifth Generation Conference. National Institute for Software.

Petrie, C. 1987. Revised Dependency-Directed Backtracking for Default Reasoning. In Proceedings of the 6th National Conference on Artificial Intelligence, 167-171. Menlo Park, Calif.: American Association for Artificial Intelligence.

Steele, R. 1987. An Expert System Application in Semicustom VLSI Design. In Proceedings of the Twenty-Fourth Design Automation Conference, 679-686. Los Alamitos, California: Computer Society Press.

Steele, R. 1988. Cell-Based VLSI Design Advice Using Default Reasoning. In Proceedings of the Rocky Mountain Conference on AI, 60-74. Denver, Colorado: RMCAI.

OHCS: Hydraulic Circuit Design Assistant

Yusei Nakashima and Tomio Baba

Uses for oil hydraulic power and control are found in almost every manufacturing plant in the construction, agriculture, transportation, automobile, marine, and aerospace industries. Oil hydraulic power systems transmit and control power through the use of pressurized oil within an enclosed circuit.

The process of designing hydraulic circuits can generally be divided into four stages: circuit design, component selection, static and dynamic analysis of the circuits, and inspection of completed drawings (figure 1). These four stages are normally entrusted to experts in respective fields. A dynamic analysis is primarily handled by the analytic experts, and the design and inspection of drawings is entrusted to senior designers.

One major problem with such a circuit design approach occurs when drawings are found that do not meet specifications, and much of the design and analysis must be redone. These additional engineering tasks cause a great loss of time and a decrease in the productivity level of the hydraulic engineer. Because so much expertise is involved in each of these engineering tasks, the traditional approach taken in developing hydraulic systems is extremely tedious and inefficient.

In analyzing the concentration of expert knowledge required for each task, we developed an expert system for hydraulic circuit design, oil hydraulic circuit simulator (OHCS). OHCS is an assistant for design

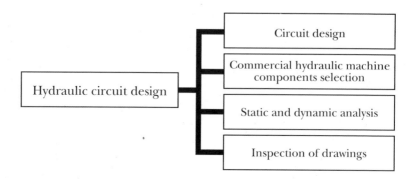

Figure 1.Hydraulic Circuit Design Tasks.

and manufacturing engineers who are responsible for efficient design and performance of oil hydraulic power systems. Our goal was to increase productivity by integrating the four processes of expert reasoning, design, analysis, and drawing inspection into a single system.

OHCS Architecture and Its Functions

OHCS runs on a Symbolics Lisp machine within the Art™ environment. OHCS consists of a Lisp program, 700 rules, and 2000 schemata. The coding ratio is seven parts Lisp and Flavors™ to three parts Art. The system requires 80 MBytes during run time for general use.

Figure 2 illustrates the architectural organization of the system, which consists of six major components. The OHCS system performs five primary functions: (1) supports the generation of circuit diagrams, (2) assists in component selection, (3) inspects diagrams, (4) generates analytic models, and (5) carries out quantitative simulation.

The Generation of Circuit Diagrams

The system is able to apply the expertise of the designers to generate circuit designs as well as computer-aided drawings (CAD) that meet specifications. Graphic symbols represent the component and all features pertinent to the circuit diagram. The graphic symbol representing hydraulic machine components is shown at the right side of the user interface in figure 3 and can be selected with a mouse. Because 150 graphic symbols represent the hydraulic machine components, selection of the various types of components is facilitated by a layered menu array. The composition of the circuit diagram, which appears in the center of the screen, is directly connected to the knowledge base,

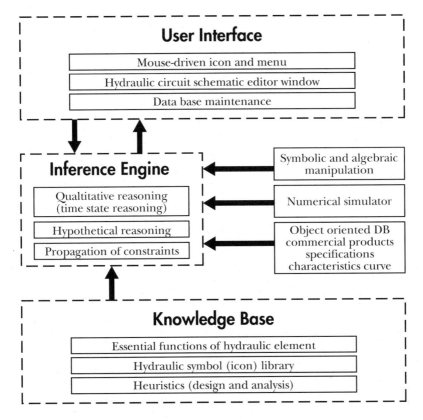

Figure 2. The Architectural Organization of the System.

and its image is recognized as knowledge (intelligent feature-based CAD). Therefore, a picture on the screen that indicates, for instance, a connection between machine A and machine B is also kept in the knowledge base as a condition in which machine A and machine B are connected. The graphic images seen on the screen are therefore not merely pictures in the ordinary sense.

Components Selection

The designer selects commercial hydraulic machine components after the generation of circuit diagrams. OHCS provides an object-oriented database, which contains the specifications and characteristics of 1500 commercial components.

OHCS automatically generates the used parts table, which is shown on the right in figure 4. This table is also the selection menu of commercial hydraulic machine components, which retrieves components from the object-orientated database.

Diagram Inspection Based on Experience. Because the system's knowledge base consists of diagram inspection expertise, the operation of diagram inspection proceeds as if such specialists were on hand to supervise the operation. In addition, the diagram inspection results, such as comments, and the parameter calculations are displayed in the upper part of the screen, as shown in figure 4, as a guide for the designer.

Traditionally, the senior designer checks to see that the functions of each circuit correspond to the specifications. This operation is about 80-percent qualitative and is deeply dependent on the special knowledge of the expert.

Automatic Generation of Analytic Models

In addition to making the hydraulic circuit diagram, simulation of static and dynamic models must be carried out. The models used in these simulations are automatically constructed in accordance with the expertise of the analysts, who express the state of hydraulic machine components in the form of ordinary differential equations. OHCS simulates the analytic expert's qualitative experience in mathematical modeling.

Quantitative Simulation

Numeric simulation of the hydraulic circuit is carried out after the automatic generation of the analytic model, which is described in terms of sets of ordinary equations. In the past, the simulation process was an operation completely separate from the process of circuit diagram construction. In OHCS, however, it is linked to the designing process, producing real-time simulation results, as shown in figure 5.

Furthermore, the simulation process is characterized as the process to be repeatedly executed with modified component parameters until designers are satisfied with the results. OHCS makes it easy to perform this process and is able to decrease the amount of time.

Circuit Knowledge Representation

OHCS uses three forms of knowledge representation: (1) a declarative knowledge base of primitives and composite components, a relation for circuit construction, a database interface, analysis model parameters for components, icon graphic symbols, and so on; (2) production rules in the form of if-then rules of many types—control of time-state reasoning and hypothetical reasoning, empirical associations of predictable problems and measures, the processing of the mouse-driven graphic

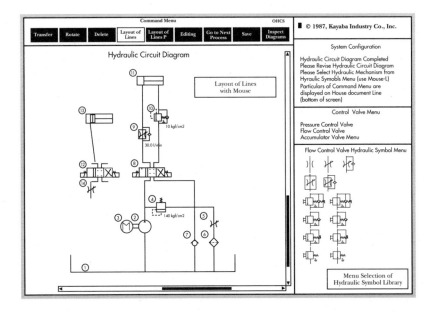

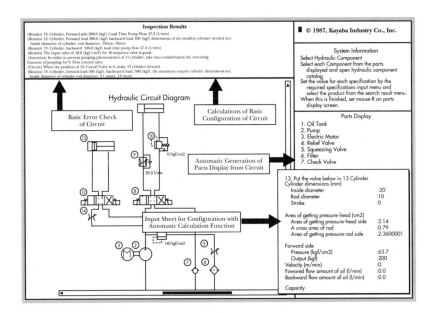

Figure 3. (above) Hydraulic Circuit Design Drawing.

Figure 4. (below) Diagram Inspection.

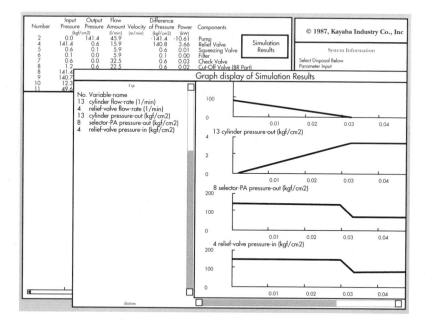

Figure 5. Simulation Results.

interface, the generation of ordinary differential equations, and so on; and (3) algorithmic knowledge expressed as functions—symbolic and algebraic manipulation to simplify and substitute equations and other calculations.

OHCS contains complete representations for hydraulic circuits, called *the circuit representation language.* This language admits a relatively simple representation using well-known ideas about schemata (frames) and inheritance. Figure 6 illustrates circuit knowledge. The graphic images seen on the screen in which the pump and relief valve are connected are expanded to the internal knowledge through relations; these are kept as part of the knowledge representation of figure 3.

Circuit Qualitative Reasoning

OHCS qualitatively identifies the circuit topology from the circuit diagram by searching the flow path and using heuristics for the following three purposes: (1) recognizing the circuit, (2) inspecting basic diagrams, and (3) giving important information for generating equations through constraint-propagation methods.

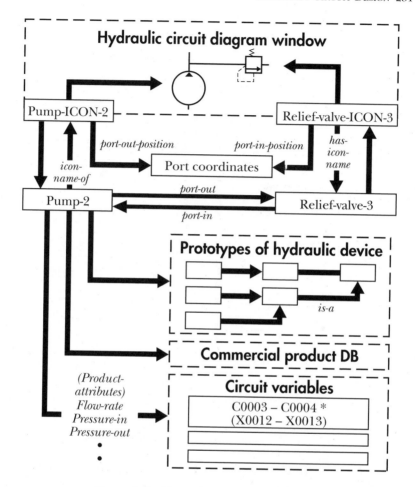

Figure 6. An Example of Circuit Knowledge.

Identification of the circuit topology is executed after piping relations, which are the port-in and port-out relations shown in figure 6, are constructed. When hydraulic machine components are added or changed by the designers, constraints with piping that exert an effect throughout the circuit system are changed at the same time because OHCS maintains the logical consistency.

At this point, OHCS infers the flow path by means of the concept of node and link. *Nodes* represent each device terminal (inlet and outlet ports of components). *Links* represent each component, which are conduits for oil that flows from node to node (Gentner 1983). OHCS uses both qualitative application of a time-state reasoning, which is

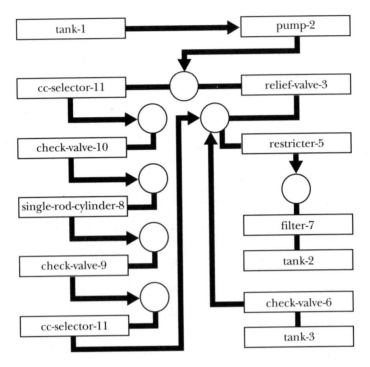

Figure 7. An Example of Circuit Topology.
Rectangles denote links (devices), circles denote nodes.

characterized as nonmonotonic reasoning, and heuristics for solving the flow-path problem.

The flow-path representation of the circuit diagram, called the circuit topology, is shown in figure 7. This is only a part of the circuit topology of figure 3.

Coupling Qualitative Reasoning and Numeric Computing

OHCS provides for piecewise linear approximations of ordinary differential equations for the circuit because ordinary circuits can be expressed as nonlinear systems. In general, piecewise linear approximations have exponential complexity in the number of nonlinear components in the circuit (Sacks 1987). Therefore, OHCS provides the search-limiting combinatorial methods, which leads to efficient analysis of the circuit with piecewise linear models.

In the search-limiting combinatorial method supplied by OHCS, OHCS generates piecewise linear equations as conflicting hypotheses

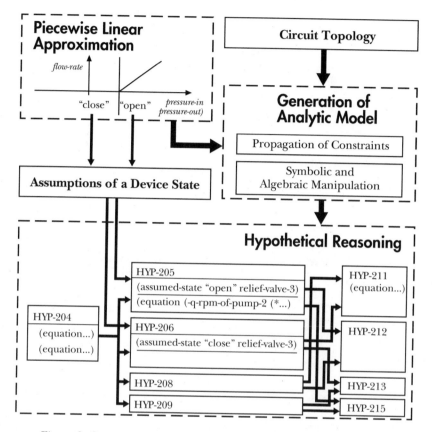

Figure 8. Generation of Analytic Model with Hypothetical Reasoning.

based on deductive logic. They can be viewed as a single analysis space for each hypothesis shown in figure 8.

While OHCS constructs sets of equations, it substitutes and simplifies equations by symbolic and algebraic manipulation, to which propagation for constraints applies (Stallman and Sussman 1977), as shown in figure 9. Instantaneously, each hypothesis is examined using both the designers' practical experience and circuit topology–based heuristics. In the course of generating hypotheses one after another and pruning them, OHCS maintains the logical consistency among the hypothetical worlds.

Consequently, the parameterized circuit equations, which are contained in several consistent analysis spaces, can be obtained. These equations are again examined using the heuristics obtained from the analytic experts. Then the equations are computed to get the circuit-

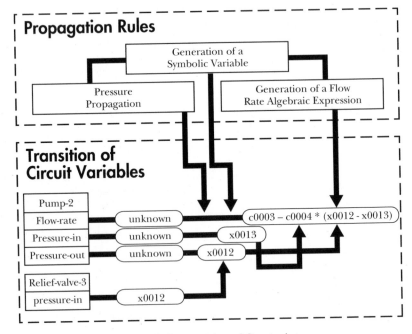

Figure 9. Propagation of Constraints.

state variables, such as input pressure, output pressure, flow rate, and heat generation.

The analysis space can to be viewed as a three-layer structure that consists of a qualitative reasoning level, a hypothetical reasoning level, and a metalevel, as shown in figure 10. Multiple contexts are expanded as network in each level. A context is a primitive in the analysis space. The metalevel contains common information through the analysis space. Information in the lower level is visible at the higher level, but the reverse is not true. Accordingly, the analytic model generator automatically generates equations by making reference to the circuit topology information.

Knowledge Acquisition

The first step in developing OHCS was to gather information from documentation and human experts. Much of the existing design documentation turned out to be defective in practice because hydraulic circuit expertise is not easily documented. As a result, only 20 percent of the documented information could be used to good effect in OHCS.

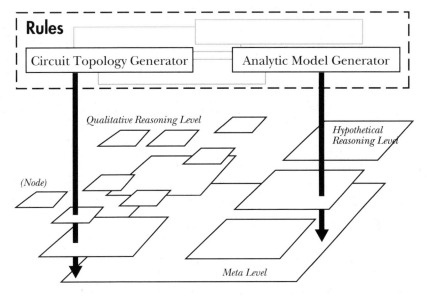

Figure 10. Multi-layered Analysis Space.

However, the documentation was important in triggering knowledge extraction when the interviews were conducted with experts.

Results

The system has been deployed since 1988. The OHCS expert system was put into operation 18 months after knowledge acquisition began; five people required for development. The payoff for our organization is as follows: (1) OHCS has reduced the amount of work at the designer level by 50 percent. If the work pertaining to analysis and diagram inspection is also considered, an even greater time savings has been achieved. (2) Maintaining the circuit diagram drawings at a specified quality level is possible because the expertise involved in design, analysis, and diagram inspection is regulated. (3) The technology-transfer process completed during the development of OHCS has further enriched analytic knowledge for the researchers who have been working on dynamic analysis.

In the general flow of design operations, the first step is the receipt of circuit specifications from the client. A table of quality requirements is then drawn up, and the actual designing is carried out. This is the area covered by OHCS. The next step is the layout design and the preparation of a quality assurance sheet. It was first thought that OHCS would fulfill a part of the general flow of design operations. In the future, we plan to build other expert systems that fulfill the remainder of the design operations.

References

Gentner, D., ed. 1983. *Mental Models.* Philadelphia: Lawrence Erlbaum.

Sacks, E. 1987. Piecewise Linear Reasoning. In Proceedings of the National Conference on Artificial Intelligence. Menlo Park, Calif.: American Association for Artificial Intelligence.

Stallman, R. M., and Sussman, G. J. 1977. Forward Reasoning and Dependency-Directed Backtracking in a System for Computer-Aided Circuit Analysis. *Artificial Intelligence.*

Media
and Music

Times: An Expert System for Media Planning
Media Top and Intellia

Wolfgang: Musical Composition
by Emotional Computation

Times: An Expert System for Media Planning

Georges Girod, Patrice Orgeas, and Patrick Landry

The television intelligent media planning expert system (Times) is an expert system for television advertising campaign design. It elaborates television planning with preserved balance and optimized performance. This planning is based on all client's data (accounts, commercials, formats, marketing targets, and periods) and television databases (audience, programs, price lists, and so on).

Times comprises several expert modules developed with NExpert, and algorithmic procedures prototyped with NExpert and rewritten in C. A rule-based skeleton drives the planning strategy and activates the tactical procedures. An original audience forecast model was incorporated into the expert system to improve the efficiency of the interactive buying procedure of the campaign screens.

Times has been successfully used since October 1988 by Mediatop. The system builds a full television campaign in four to eight minutes (instead of three days for a human expert). The results are better than those achieved by human experts.

High-quality planning, reliability, flexibility, and savings in time and money were immediate benefits for Mediatop. An even more profitable consequence of the expert system will be the commercial benefit to Mediatop's reputation in the use of high technology in media planning.

This article is divided into three sections:
1. The expert's point of view: Why Times was developed
2. The technical point of view: How it works
3. The user's point of view: What the performances, major benefits, and future product improvements are

Why Times Was Made:
The Activity of Media Planning

Times is an expert system dedicated to improving media consultant proposals in TV planning. *Mediaplanning* can be considered the technical branch of the advertising profession. When a company wants to advertise its activity or products, they ask an advertising agency to design a campaign . This campaign has a specific message to deliver to a selected target (for example, men from 20 to 45 with a high income) with maximum efficiency. The media-person working with the ad-person is in charge of selecting the best media to accomplish this goal.

Television Planning

Television planning is crucial. The client, of course, wants the best possible results on the investment. The choice of the medium is usually fairly easy, and for mass-market products, television is most frequently selected: No other medium has its phenomenal reach (that is, percentage of the target exposed to the message) and impact. Nevertheless, even if television has the best cost-efficiency ratio, the average cost for a television *flight* (a list of spot screens, roughly from 10 to 100, each one selected for specific criteria: cost efficiency, reach, affinity with the target, and so on) in France is about $1 million and can easily reach $10 million (and five times more in the United States). Moreover, to buy a single 30-second spot during prime-time costs about $100,000 in France. Because the actual number of people watching during prime time can vary as much as fifty percent (and even more if the target is specific), the choice of the media-planning agency is of the utmost importance.

The Information Processed

To design a television flight, media planners refer to objective data about the screens, such as audience per target, price, stability of the audience, and affinity with the target group, as well as subjective data such as television programs (what is the impact of changing the evening news announcer?), and qualitative environment (is it convenient to advertise my product in this context?).

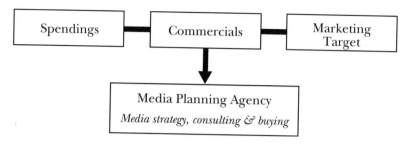

Figure 1. TV Media Planning.

The complexity of this process can be compared to playing chess. Selecting one screen is like moving a pawn: The move is insignificant by itself; it is only important as part of a whole strategy. Also, no ultimate answer exists about what is good or not: Any move can be interesting in certain circumstances.

The Software Tools

Until now, media planners only had basic software—capable of adding prices and estimating the overall performances and balance between reach and frequency defined as the average number of contacts—to enable them to deal with all this information. None of them were able to design a television campaign using this limited software. To obtain a television plan required three days of intensive work and justifications about how and why it was built that way were fairly inaccurate.

The Expertise

Nevertheless, the expertise clearly existed because some media planners produced excellent results: In several countries, their salaries are linked to their performances. Thus, all the ingredients existed for developing a media-planning expert system: the complexity of the problem; quantitative and qualitative data; economic reasons; the easily measurable impact of the choices; and, last but not least, the expertise itself.

Several efforts were made to develop such an expert system in France, but until now these attempts were unsuccessful. The development of Times was a formidable challenge, and its answer to quality media planning should be considered a major improvement in advertising.

AI Methodology: Analyzing the Expertise

Building a television campaign is typically a highly expert task, dealing with an impressive volume of information: Television data (programs,

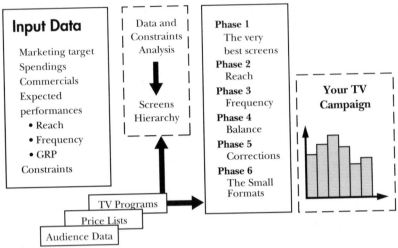

Figure 2. AI Methodology: Architecture.

audience, price lists, screen availability, commercial negotiations with the channels, and so on) and client data (expenditures, market data, target, commercials, wishes and expectations, and so on). Moreover, the financial challenge is important. Increasing efficiency by 10 percent on a $1 million television campaign can be rewarding.

When we first began working on the Times project, we faced a typical AI situation: Expertise that was particularly complex, and the poor initial ability of the expert to explain how he elaborated a television plan. Intellia's knowledge engineers adopted an original methodology that proved successful. The expert was first trained on NExpert for two weeks and then learned to prototype small sub-problems with Intellia's knowledge engineer for another two weeks. At the same time, the expert was taught self-analysis techniques so he could model his own expertise within a five-month period. The knowledge engineer's involvement was reduced to one half-day meeting per week for expert model control and project guidance. The result of this analysis was an expert's full knowledge that could be exploited and later formalized in natural language description, rule-based prototypes, and symbolic algorithms.

The success of this technique resulted from the expert's ability to adopt different mental schemes and methods. The resulting benefit was a deeper understanding of his own expert knowledge and an accurate planning strategy.

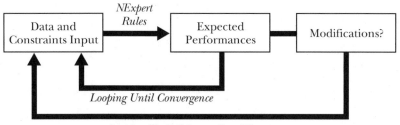

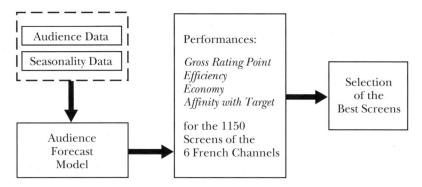

Figure 3. (above) Data and Constraints Input.
Figure 4.(below) Screens Analysis and Sequences.

Architecture

The main architecture of the system consists of three modules. The first module is an input data form. The second module performs a data analysis task, and generates a screen hierarchy. The third module makes the planning design in six phases.

The Input Data Module

The input data module acts as an intelligent data filter. Spendings, marketing targets, commercial characteristics, campaign period, price list modifications, and so on, are entered so that any data input or modification triggers calculations of the expected performances. These calculations are gross rating point, coverage, frequency, and number of spots related to the constraints.

These average performances can be chosen as default values by the operator, or the user can force the expected performances to influence the planning strategy of Times. For instance, a high number of spots tar-

Strategy: a NExpert set of Rules

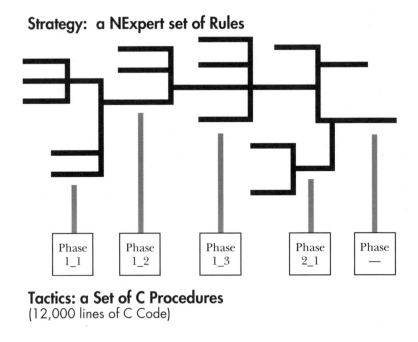

Tactics: a Set of C Procedures
(12,000 lines of C Code)

Figure 5. Building the Plan: Strategy and Tactics.

get the system toward low-price screens and favor the day-time and night-time screens instead of prime time. However, Times never feels obliged to satisfy contradictory data or wishes; these expected performances are always considered as planning strategy indications that Times preferably tries to satisfy. One of the most pleasant features of Times is its refusal to spend all your budget if it thinks it unnecessary.

The input form module is designed as a double loop: A NExpert set of rules evaluates the estimated performances until convergence of the calculations, and the evaluations can modify some of the user's wishes (because they contradict other wishes). The process loops until validated by the operator.

The Screens Analysis Module

This module begins with a classic media-planning computation. From audience and seasonal data, a mathematical audience forecast model evaluates the performances of the spot screens. Thus, for the 1,150 available screens on the six channels of the French television market, the gross rating point, efficiency, economy, and affinity with target for each spot screen are calculated.

If	PHASE_4_4_0	is TRUE
	PLAN.STRATEGY	is "delicate"
	DAY_TIME.SCREENS_RATIO	<= coeff_7
	GRP_CORRECTION	<= 0.23
Then	PHASE_5_0	
And	CORRECTION	is set to DAY_TIME
	RESELLING.STRATEGY	is set to NIGHT_TIME
	BUYING.STRATEGY	is set to DAY_TIME
	ORDERING	is set to META_RANK
	EXECUTE	PROCEDURE_19 @ATOMID= 12,14,3,18;

Figure 6. Example of Strategy Rule.

A set of rules then analyzes these performances and deduces their nonlinear combination, which allows the construction of a power function for each screen, and main hierarchy of these screens in decreasing order of the power function.

This hierarchy is fundamental for television planning design because the system constructs the television campaign by choosing primarily among the best 30 percent of this list.

Building the Plan: Strategy and Tactics

The reasoning of the media-planning expert is a combination of strategy (what goals should be focused on at this stage of the plan?) and tactics (what is the best spot screen—or subset of screens—to achieve the immediate goal?). This reasoning process is comparable to that of the chess player's; the number of possible combinations is tremendous in both cases! The technical solution for managing this double reasoning process was to build the strategy with a NExpert skeleton of rules driving a set of 29 main tactical algorithmic procedures.

Strategy Rules. These procedures are triggered by rule reasoning, which determines what goals are to be reached, at what time, and with what means. Figure 5 presents an example of strategy rule.

Altogether, the procedures represent an impressive 12,000 lines of C code.

Tactical Procedures. Figure 6 presents an example of a simple tactical procedure.

This reasoning was first prototyped with NExpert, using non-monotonic logic and reset actions to obtain adequate looping until the desired goals are reached. The first advantage of this prototyping method

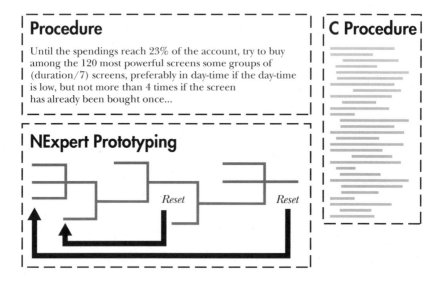

Procedure

Until the spendings reach 23% of the account, try to buy among the 120 most powerful screens some groups of (duration/7) screens, preferably in day-time if the day-time is low, but not more than 4 times if the screen has already been bought once...

C Procedure

NExpert Prototyping

Reset   *Reset*

Figure 7. Example of Tactical Procedure.

is the accuracy and flexibility of the expert reasoning model. Moreover, the prototype set of rules is used as an algorithmic model for efficient and fast coding.

This example illustrates the use of an expert system shell as a case tool, and obviously, NExpert's ability to graphically visualize rule chaining was an overwhelming advantage.

Backtracking. Times reasoning is a continuous back-and-forth process; the system has to keep each step in mind to avoid looping among unattainable goals or contradictory requirements.

The backtracking in Times is controlled by both the nonmonotonic logics of NExpert (resetting a reasoning phase or sub-phase with limited depth) and the algorithmic procedures control (while and for loops submitted to control variables). This technical point was one of the key features during validation. (Of course, we do not pretend here to perform as complete a backtracking guidance as must be done in operational research.)

Audience Forecast Model

As can be expected, the prime-time screens (from 7:30 P.M. to 10:30 P.M.) have the highest audience as well as the highest price. The capacity to provide a better audience forecast on these prime-time screens

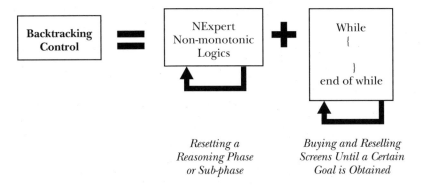

Resetting a   Buying and Reselling
Reasoning Phase   Screens Until a Certain
or Sub-phase   Goal is Obtained

Figure 8. Backtracking.

and, thus, optimize clients' spending is a decisive factor in television planning strategy.

Times contains a forecast model (today restricted to prime-time movies) based on a classical mathematical model (computing tendency and seasonably) and a rule-based model that gives a context-driven correction of the mathematical forecast.

For example, Times forecasts the audience of *Gone with the Wind* on Channel 2 at 23 percent, given the following criteria: This movie is an all-time classic, featuring well-known players; it is more than 10 years old, is dubbed in French, and was not presented on television during the last two years. Furthermore, Sunday evening is a popular viewing time, and the competition on the other channels is poor on the scheduled evening.

This feature of Times is a first step toward the availability of a reliable audience forecast model. Today, everyone recognizes the limits inherent in classical models, which have not evolved significantly over the last 15 years. We are convinced that these limitations can be overcome with AI techniques and accurate expertise.

The Exploitation of Times

After several months of intensive use, Times was a fully operational system. Its first and most exciting result was the satisfaction of the client, the media-planning agency, and the expert.

Performances

Time's performances are satisfactory. Concerning rapidity, a full session, from input to the proposal and its analysis, only takes four to

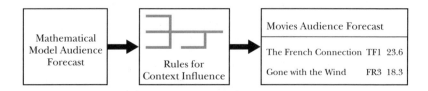

Figure 9. An Expert Module for Prime-time Movies Audience Forecast.

eight minutes, instead of the few days required when done manually. Concerning quality, the accuracy of the result is extremely high, not only from the expert's point of view, but also from that of independent media planners. The reason for this high accuracy is the extremely precise procedure Times follows to reach its final proposal. The expert system makes as many as 3 million screens analyses with all the expert's rules—an impossible for any human expert!

Benefits

These performances mean four important benefits for the user:

1. Efficiency and time savings. The everyday use of the system is so easy that any assistant can use the knowledge. After a quick demonstration of how the system works, all Mediatops staff members have been able to use Times without prior training. Moreover, Times has proved to be a good educational tool. The justifications for choosing any single screen are explained through a real-time reasoning tracking. This first benefit is both qualitative and quantitative (about a 40 percent time savings for the expert).

2. Improved client service (better planning and, specifically, real-time satisfaction): Any modification in one single input datum can significantly modify the agency's proposal; in a fast moving area such as television, this is highly significant. Until Times was available, no one would have seriously considered replanning a whole flight for a minor datum change; modifications were just added to the preexisting campaign.

3. Overall quality improvement. The time and mental energy saved by the expert can be devoted to improving theory and methods. As in any business, media experts spend too much time doing day-to-day work, with no real opportunity to question their knowledge. Times has made the situation different. Whenever the expert system comes up with an unexpected solution, the expert can (and has to) examine its reasoning to understand the situation and modify the expertise if a disagreement exists. This situation is a permanent challenge for the media planner.

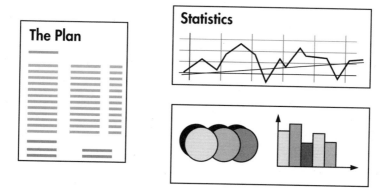

Figure 10. The result.

4. Commercial development. Times is such an innovation to media planning that its impact can be decisive in convincing new clients. The expert system can be a forceful way to illustrate the commitment to improving media-planning performances. In this way, Times is a powerful communication tool.

Conclusion

Because of its novelty, Times will remain an exclusive product for the company for which it was originally designed. Nevertheless, its success is strong enough for the product to be adapted to an endless variety of applications. Its second implementation will be for a French television channel. The market for this product is worldwide, even if some specifics have to be adapted to fit other national needs. Television was certainly the most interesting and difficult media to serve, but other media, such as radio, press, billboards, and multimedia can be used.

Another derivative expert system application should be quality audience forecast. Today's mathematical models have reached their limitations, and the expert system technology will introduce qualitative and context data into the forecast model. The opportunities following the success of Times are so interesting that both Intellia and Mediatop have created a joint venture subsidiary.

Wolfgang: Musical Composition by Emotional Computation

R. Douglas Riecken

A man, viewed as a behaving system, is quite simple. The apparent complexity of his behavior over time is largely a reflection of the complexity of the environment in which he finds himself.

—Herbert A. Simon

Processes such as creative thought and learning result from achieving a structured pattern of goals. These goals evolve as artifacts from the environment within which a system exists. As an example, consider an infant's ability to navigate around a physical object, such as a sofa, while moving from a source location to a destination location. The goal of getting around a physical object thus, becomes an artifact of the infant's movement within its environment. Historically, artificial systems have also depended on specific sets of goals to direct learning and problem solving within a given domain. A major difference between the infant and a mature artificial system is the higher degree of complexity by which the infant selects goals. The biological and environmental heritage of human cognition favors construction of metagoals that satisfy given emotional states. These metagoals compile sets of problem-solving goals, which comprise partial mental states that the human system feels a goodness for in the context of its current emotional metagoal. These partial mental states are reconstructed from previous total mental states of memorable events. The selection of memorable events is based on the known constraints of the current goal-problem. (The selection process is guided by search strategies such as analogy, reformulation, heuristics, and association.)

The application presented in this chapter, a system called Wolfgang, began as a prototype examining several methods of goal formulation to artificially compose music. The principal problem in the design of Wolfgang was defining a successful theory of goal formulation to support system evolution. The evolution of an intelligent system requires its perception of environmental events and phenomena and their transformation into useful knowledge, thus improving the system's performance within its environment. The system must be capable of determining what goals best classify the perception of an event and its coding in memory.

The theory of goal formulation used to design and implement Wolfgang is based on the premise that memory provides the ability to recreate partial states of mind and that the selection of these states is determined by the current disposition of the system. Thus, a system formulates goals associated with specific emotional qualities that satisfy its disposition. The theory assumes that all systems are conceived with an innate set of primitives, some form of genetic code that defines the evolution of each individual system. These primitives propagate in a defined manner throughout the entire system, controlling the behavior of the system and the perception of its environment.

K-lines and E-Nodes

Wolfgang identifies two foci of interest: (1), a theory of goal formulation directed by emotional constraints and (2) a new method to artificially compose music. Over the past two decades, a considerable amount of work has successfully addressed the domain of computer-composed music. This work has resulted in many syntax-oriented theories and has raised an important question: Do such systems truly compose music, or are they really music generators? Although it is essential to compose music by some well-formed theories of grammar (Lerdahl and Jackendoff 1983; Piston 1941), these theories fail to capture the essence of music, its ability to communicate emotion—the semantics of musical sentences. Wolfgang's use of emotional computation, K-lines, and musical grammars presents a new direction for composition by computer.

A principal influence in the design of Wolfgang is Marvin Minsky's *society of mind theory* (The society of mind theory is the evolution of Minsky's *K-line theory* [Minsky 1986]). The theory describes the behavior and cognition of a system by the dynamic relationships of linked semi-intelligent agents called K-lines. The theory goes on to explain that sets of K-lines form societies of agents providing specific mental functions and that these societies or agents can dynamically form multiple connections (K-lines) with other societies or agents to provide many different types of intelligence. It is the activation of sets of K-lines that compose partial mental states. A total mental state comprises several partial mental states active at a single moment in time. Further, system evolution is supported by constructing new K-line connections within or between these partial mental states.

A key component in Wolfgang's architecture is a network of K-lines. The composition of the K-line network consists of many heterarchical societies of agents. Eventually, these societies decompose to individual agents, which, in turn decompose to an innate set of primitives that defined Wolfgang at its inception. (The inception of Wolfgang is that moment in time before the system began to perceive its environment and learn.) These primitives, or *emotional nodes* (E-Nodes), are fundamental structures that qualify the emotional characteristics in perceiving the simplest elements of a given domain or environment.

For example, in the domain of musical composition, an E-node defines the emotional qualities of such auditory stimuli as musical modality, single harmonic structures, harmonic progression, amplitude, tempo, rhythm, and musical intervals. Metaphorically, E-nodes serve as the system's emotional genetic code.

Each E-node contains a set of real number weights ranging from zero through one for each of the following emotions: happy, sad, anger, and soul searching. These weights define the emotional character of a given stimuli and are constant for the life of the system. The E-nodes within Wolfgang serve as a lexicon defining the emotional characteristics of individual agents in the K-line network. Thus, as Wolfgang evolves, the properties of the E-nodes propagate throughout the entire system, contributing to the perception of environmental stimuli and the creation of K-lines to represent more complex societies of musical agents with their respective emotional weights.

If E-nodes define the emotional capacity of such musical atoms as tempo, then what phenomenon justifies an E-node? Psychological research in music addresses the extent to which the construction of a melody is determined by factors extrinsic to music, in particular, the natural pitches of the harmonic series and their effects on and within the ear. For example, musical idioms of the world as a whole demonstrate a high salience for the octave, perfect fifth, and major triad because of the harmonic series (Jackendoff 1987; Balzano 1982; Clynes and Milsum1970; Clynes 1977). Thus, the idea is to define a set of E-nodes that are innate throughout all cultures of music.

As an example, repertoire for the kyoto, sitar, balalaika, lute, and guitar all demonstrate a high salience for octaves, perfect fifths, perfect fourths, cycles of fifths, and cadences built from perfect fifths. These musical components are strong and support a sense of finality. Thus, they are useful in constructing resolution from tension or climax. Other components such as major and minor seconds, minor thirds, augmented fourths, minor sixths, and minor sevenths are less stable and are useful in generating the tension or climax.

Time trajectories are another important property in music perception. Faster tempos support higher levels of energy; slower tempos perform the inverse (Minsky 1986; Clynes 1977). Another consideration is the spatial proximity of pitches in a melodic sequence and its role in the effectiveness of memory to retain the specific melodic sequence. Studies have shown that smaller-sized intervals are processed more effectively (Deutsch 1982a, 1982b). This consideration is important for motivic development.

Application Overview

The discipline of musical composition requires balancing in parallel the goals of many distinct musical components (for example, melody, rhythm, harmonic progression, motivic development, and form) that contribute to the highly complex structure of a musical work. Each goal considers the structural, aesthetic, and functional constraints of its respective musical component based on how well-formed a musical grammar is for each component. Wolfgang was designed to formulate goals and subgoals to meet the respective constraints of each musical component and to formulate metagoals to negotiate the collaboration of these diverse component goals during the composition process.

A typical session with the application allows the user to request Wolfgang to compose a musical composition that communicates one of the specified emotions. Currently, the musical components that contribute to the emotional character of a given composition composed by Wolfgang are melody, harmonic progression, rhythm, tempo, and motivic development. To limit the class of problems supported, the length, meter, and form for all compositions are static and serve as a skeletal structure for the planning functions in Wolfgang.

Each composition is exactly sixty-four measures in length with a meter of four-four time and follows a quasi Sonata Allegro musical form. The

Sonata Allegro form is partitioned into three sections: an exposition section of 32 measures with a modulation, a development section of 16 measures with a modulation back to the original key or modality, and a recapitulation section comprised of the final 16 measures in the original key. A detailed view of this skeletal structure follows:

The first 16 measures of the exposition section state the entire motif (theme) of the composition and then develop the motif or subsections of the section with slight variations. The second 16 measures of this section are dedicated to pure development of the motif. The methods of development during this 16-measure section are allowed greater freedom. This section then concludes with a modulation to another key or modality.

The development section, the next 16 measures, continues the development of the motif in the new key or modality with the same amount of freedom allowed in the second half of the exposition section. The development section then closes with a modulation to the original key or modality of the exposition section.

The recapitulation section begins by restating the original motif. The remainder of the recapitulation is then composed of several of the most memorable motif-developed artifacts from the exposition section. This is accomplished as follows: During a composition session, feedback facilities allow the application to monitor its work. As phrases of musical sentences are constructed, they are weighted by their success in satisfying the current disposition of the system. Thus, once the original motif is restated in the opening measures of the recapitulation, Wolfgang cuts and pastes the musical phrases with the highest weight values from the exposition to complete the composition.

The physical location of cadences is every four or eight measures depending on the length of the motif (which is two or four measures). The system then maintains complete symmetry of musical phrases by holding a constant distance between cadences throughout the composition. From a user perspective, a typical composition session with Wolfgang is partitioned into three phases: priming, evaluation, and composition.

During priming the user performs three tasks: (1) enters a *seed* musical motif (a linear set of musical tones and the duration of time for each tone), (2) selects the emotional quality that the composition should communicate, and (3) defines the behavior of Wolfgang during the composition session as a type ranging from conservative to aggressive. The behavior of the system during a session directs goal formulation to try ideas and methods ranging from conservative and established to aggressive and new.

During evaluation Wolfgang evaluates the user-supplied motif to determine a characterization of the motif that best matches the desired emotion of the completed composition. This determination entails parsing the motif into simple elements consisting of several notes each. The selection of these elements is based on a melodic grammar and the emotional characteristics of specific groupings of notes that best communicate the emotion selected by the user. Wolfgang then examines the combinatorics of harmonic progression for the parsed motif and defines several options of harmonic progression implied by the motif that capture the desired emotional qualities.

During composition Wolfgang composes the musical work.

System Architecture

Wolfgang's architecture is composed of the following elements: (1) a K-line network, (2) a scheduler, (3) a blackboard, (4) several feedback loops, (5) a log file, and (6) a user interface.

K-Line Network

A K-line network can be viewed as a frame-based semantic association network. The K-line network is partitioned into two parts, with each part providing a distinct function. One portion of the net, the methods net, serves as a composer's tool box. The methods net contains several different method societies of agents defining such compositional methods as cadence, transposition, and motivic development. It is important to note that the methods net is composed of facts, not rules. The other portion of the K-line net, the facts net, explicitly defines such musical facts and artifacts as sets of intervals, rhythmic patterns and harmonic progressions.

Each method society in the methods net is constructed by interconnecting many smaller societies or individual agents defining specific types of methods. For example, the method society defining motivic development is composed of many smaller societies defining types of development, such as inversion, retrograde, and elongation. K-lines interconnect the method agent with each of its type agents. In turn, each type agent is represented by one or more K-lines as specific musical facts or artifacts in the facts net. These K-lines to the facts net define an explicit musical artifact, such as an exact set of intervals to compose a musical phrase to satisfy the respective type agent, which in turn satisfies the respective methods agent, which in turn satisfies the system disposition.

One major difference exists between agents in the methods net and the facts net. Agents in the facts net are assigned emotional weights when they are created. The agents in the methods net only temporarily inherit emotional weight values. When a given set of K-lines are active, the agents in the methods nets temporarily inherit the weights of the explicit musical artifact most active in the facts net they are connected to.

Implementation of K-Lines and E-nodes. Within Wolfgang, E-nodes are implemented as frames. The first slot in the frame explicitly denotes the type of musical component or stimulus. For example, the component could be a specific tempo setting or the interval between two pitches. The next four slots hold the four respective emotional weights. These weights are assigned at the inception of the system and remain constant during its life. Each E-node maintains all four emotional weights because a given stimulus might be capable of communicating several different emotions. Each E-node is then associated with a specific group of E-nodes based on its respective function. (Examples of group functions are tempo, intervals of two pitches, and modality.) These E-node groups are implemented as lists.

It is important to remember that an agent with a set of data links is a K-line and that the set of links interconnect sufficient data to allow the K-line to represent a useful fact of knowledge. Each K-line results from constructing connections for one of the following cases: (1) between sets of E-nodes, (2) between E-nodes and other agents or societies, or (3) between sets of agents or societies.

Each K-line agent is implemented as a frame, known as a K-line frame (KF), and is stored as a slot value in another frame called a musical component-frame (MCF). The first slot value in MCF denotes the type of musical component. The remaining slots in MCF are called KF slots.

Each KF slot defines a given K-line associated with the respective musical component. The number of KF slots is dynamic because new K-lines are always possible. The first slot in KF defines the respective K-line. All the respective links that compose the K-line are stored as a list in this first slot.

The remaining four slots in KF denote the four emotional weights provided by the respective K-line. Each emotional weight is a tuple, where the first variable is a quasistatic weight (QW), and the second is a dynamic weight (DW). These weights are used to control the firing of K-lines. When a K-line is first constructed, each emotional weight is computed and assigned to both variables in the tuple. Two cases need to be considered when computing emotional weights: First, if a K-line connects only one E-node frame with MCF, then the respective KF weights are inherited from the E-node, and the K-line is considered a low-level agent. Second, if the K-line is connected to several E-nodes or other agents or societies, then the averaged weight of all the connections for each type of emotion is assigned to each of the respective KF weights. If a majority of the connections are to E-nodes, then the K-line is considered a low-level agent.

Each time the system is initialized for a composition session, DW of each tuple is assigned the value of QW. The DW is used in determining which K-line offers the best solution to satisfy a given emotional metagoal.

During a session, DW will change based on events in the K-line net; for example, the current system goal is to fire the K-line with the highest sad DW, which addresses the musical syntax of the current problem. Once this K-line has fired, its sad DW is decremented. This allows other K-lines a chance to fire. Eventually, the sad DW of this K-line is incremented back to the value of QW.

The incrementing of this DW occurs each time the same type of problem is addressed. Of course, the system's feedback facilities can flag the use of this K-line as a positive event and strongly recommend to the system scheduler that this K-line be fired again to address this type of problem the next time it occurs. The scheduler then assigns the value of QW to DW.

QW serves as a control weight throughout a composition session, but over time (many sessions), QW might change. The system maintains a log file of the last 20 sessions. If a given K-line shows a pattern of high or low use, then the system readjusts QW accordingly, thus allowing the system to adjust its musical opinion to evolve.

Scheduler

The scheduler conducts the flow of events within the system so that they satisfy the system's current disposition. This conduct implies that the disposition might change during a composition session. As an example, assume Wolfgang is composing a happy composition. For the first 40 measures of the composition, Wolfgang's disposition is to compose happy, but a K-line in the method net that defines the need for a small change in the emotion of the music just fired. In satisfying this new goal, Wolfgang's disposition is to compose the next eight measures of the composition as slightly angry, then change its disposition back to happy for the remaining measures of the composition.

To satisfy the system's disposition, the scheduler constructs goals, evaluates the network, activates K-lines, builds K-lines, computes and adjusts weight values, monitors feedback loops and the log file, and manages global data stored on the blackboard. Because the scheduler is responsible for so many functions, it's no surprise that a composition session runs 8 to 10 minutes. Wolfgang is currently a stand-alone system, but its modular architecture provides for future integration with other systems to distribute the processing of a session and add more features. This application is an excellent candidate for parallel hardware because of the parallel nature of the K-line network.

Blackboard

The blackboard provides global information of all concurrent and parallel events within the system. Output from evaluating the user-supplied motif and the actual composition itself are stored on the blackboard. This information is stored in several interconnected structures.

As an example, I present the internal blackboard representation of the user-supplied motif. The pitches and rests from a user-supplied motif are assigned as leaves in a time span tree (TST), as described by Lerdahl and Jackendoff (1983). Important submotifs that support the desired emotion of the completed work are marked as useful elements for the composition process. The location of each marked leaf is added to a list of good musical ideas for motivic development. Thus, this motivic element list (MEL) is a list of links into TST.

The final structure, the list of harmonic links (LHL), is a list of nested lists. The outermost list defines multiple harmonic progressions that support the entire motif and the desired emotion of the composition. Each expression in the list defines a specific harmony, which is linked to a specific leaf in TST. When multiple harmonies support the same leaf, then the respective expression for this leaf returns a list of expressions when evaluated, with each expression defining a valid harmony for the respective leaf.

The output from the composition session is stored in the same manner. The blackboard also stores output from the feedback loops, the current disposition of the system, the desired emotion type for the composition, and path names of activated K-lines in the network.

Feedback Loops

The feedback loops provide Wolfgang with the ability to self-evaluate the artifacts of its performance during a composition session. During the composition session, Wolfgang can mark for future use those artifacts that compute high emotional scores, while it satisfies the current disposition of the system. The feedback loops are implemented as background processes which send status messages to the scheduler that an interesting event has occurred and then post the blackboard with data concerning the event. The feedback loops monitor both the session's output on the blackboard and the log file.

Logfile

Although Wolfgang's compositional style has a distinct signature, the log file ensures that each composition is unique. The log file provides information of previously composed works; thus, Wolfgang can avoid the excessive repetition of musical artifacts by reviewing its composing habits during a session. The log file consists of trace data from the last 20 sessions.

User-Interface

The user interface supports interactive learning exercises for the user to instruct Wolfgang, the priming of a composition session, the tracing of the composition process, and the generation of the finished musical score. Currently, the interface is a simple interactive shell to enter commands, but future work should improve this facility.

Goal Formulation

During a composition session, goal formulation is first directed by the context of the problem. Further pruning is achieved by defining the desired emotional characteristic that satisfies the current disposition of the system. The result is the formulation of a goal to construct a set of musical component agents that satisfy the prescribed emotional characteristic. Once a goal is defined, those K-lines are fired that best match its syntax and are weighted the highest for the current desired emotion meeting the current system disposition. This process repeats recursively for each subgoal until the artifact (musical component agent) of the respective subgoal is determined to be an atom.

To demonstrate the goal-formulation process, let us examine the following example of system events. Assume that the system is about to compose the next musical phrase of the composition beginning at measure 17 (second half of the exposition). The user has (1) entered a 4-measure motif, (2) requested a happy composition, and (3) stated that the system's behavior is conservative. Within the K-line net, the method net has fired K-lines noting the cadence of the last musical phrase. Also, the current disposition of the system is (still) happy. Now the stage is set.

The firing of the K-line noting the last cadence invokes a planning goal for musical form (non-emotional goal) to determine what to do next. The scheduler determines from TST of the blackboard's current composition that the next musical phrase begins the second half of the exposition, which calls for more complex motivic development. The scheduler then constructs a goal to determine how to develop the motif at this temporal point in the session or composition by firing the method net K-line for motivic development.

The firing of this K-line propagates other subgoals, such as (1) How long will the next phrase be? (2) How should the next phrase be bridged from the cadence of the previous phrase? (3) What type of motivic development is best? I simplify this example by only examining the selection of the motivic development type (MDT).

To select MDT, the scheduler must be aware of the current temporal moment in the composition; for example, if a cadence is near at hand, then the options are limited. The scheduler must also pay close attention to the preceding phrase and cadence to maintain a smooth transition between phrases. Because in this example, we are at the beginning of a phrase, we are not limited by a cadence within the next measure or so. As a result of these findings and the firing of the MDT K-line, all K-lines for a specific type of motivic development (STMD) connected to the active MDT in the method net are considered. Remember, one of our current goals is the type of motivic development to use.

The scheduler reviews each STMD K-line with its emotional weight while determining which submotif from MEL on the blackboard to use for the development. The scheduler then fires the STMD with the highest weight that satisfies the system disposition to compose happy.

Let us assume that the selected STMD calls for repetition of the sub-motif selected from MEL but with an alteration of the interval between the first and second notes in the submotif. The firing of this K-line defines a new goal: What interval should be used? This new goal is satisfied by firing one or more K-lines in the facts net. The interval selection is dependent on many other parallel events (other active K-lines) for such societies as harmonic progression and rhythm. Let us assume for the sake of simplicity that a specific harmonic progression and rhythm were defined for this current musical phrase. Once the goal to define a specific interval is instantiated by the firing of the respective STMD K-line, multiple K-lines defining specific intervals in the facts net are considered by the scheduler.

These specific K-lines are selected by syntax first: Find intervals that slightly change the original interval of the submotif and that complement the selected harmony and rhythm. From this set of K-lines, the scheduler then fires that K-line with the highest weight matching the current system disposition. By firing this respective K-line for a specific interval, which is also defined as an E-node, we reach an atom; thus, recursion ends. It should be noted that if the K-line search fails to satisfy a given goal, then backtracking occurs.

I want to point out that the recursive K-line paths activated in this example are built from several different partial mental states. The partial mental states extracted from previous total mental states of memorable experiences are what Minsky defines as levels, level bands, and fringes in his SOM theory.

System Evolution

The agents for each type of musical component are the result of environmental stimuli during Wolfgang's evolution. These stimuli are perceived or learned and stored as the memorable events (agents) of a given mental state. The construction of mental states, both partial and total, is based on Minsky's (1979) K-line theory of memory. Further, during the perception-learning process, the memory encoding of memorable events includes the evaluation and assignment of sets of emotional weights defining the emotional qualities of the respective memorable events. This provides a facility to compute the emotional qualities of different partial mental states composed of given agents. The evaluation and assignment function of emotional weights to memorable events (agents) is computed from the respective E-node and K-line emotional weights of the current mental state.

System evolution occurs whenever K-lines are created or altered within Wolfgang. These changes in the K-line network result from either inductive learning sessions with a teacher or by experiment and discovery during a composition session.

During an inductive learning session, a teacher presents a learning example and directs its use by the system (Michalski and Chilausky 1980; Michalski 1982). Once the example is learned, the system's mental state of this memorable event results in a change in the K-line network, thereby completing the learning from the instruction phase of the inductive learning process. Using generalization rules of musical grammars, Wolfgang then attempts to create or alter K-lines within the network to extend the scope of generalized concepts constructed from the learning example to include other instances.

Learning by discovery is accomplished by formulating analogies that serve to connect weakly or never connected artifacts to determine new useful results (Guha and Lenat 1988; Lenat 1982). Discovery learning occurs when Wolfgang's behavior is selected by the user to be aggressive during a composition session. Wolfgang experiments by formulating analogies to connect weakly or never connected K-lines (musical agents) to determine their usefulness. When discovery learning occurs during a composition session, a K-line is created and emotionally weighted, defining the connection. The new agents from this K-line are then used in the respective portion of the musical work. A flag is set to mark the use of these new agents and requires Wolfgang to evaluate their impact during the remainder of the composition session. Should a negative impact occur, Wolfgang backtracks, purges the newly created K-line, and attempts a more conservative solution; otherwise, the K-line is retained in the network because of a positive discovery.

Correctness

The correctness of Wolfgang's computation based on disposition is dependent on the correct propagation of emotional weights within the K-line network. Evaluation of the correctness required the creation and evolution of several individual (Wolfgang) systems. Each system followed a distinct evolutionary path of learning exercises and composing sessions. The objective was to examine the growth of K-lines within each system and monitor the trace files during the composing sessions of each system.

The examination of K-lines revealed that their emotional weights correctly characterized their respective emotions based on the emotional genetic code defined in the E-nodes. The K-line examination evaluated the creation and altering of K-lines and the summing and averaging of weights to define individual agents and societies of agents within the network.

The trace files record the sequence of actions performed by Wolfgang to formulate goals to satisfy a given disposition. The examination of trace files served to determine if Wolfgang had formulated goals expressing a given emotion that did not match its disposition. Did the achievement of a goal result in composing a sad musical artifact while Wolfgang believed the artifact communicated happiness? The data collected in the trace files demonstrated that Wolfgang had successfully formulated goals to match dispositions.

A second evaluation of Wolfgang is its pragmatic use. After a year of prototyping, several systems have been deployed to complement the work of several (human) composers.

A total of 18 months went into the prototyping and deployment of Wolfgang. Wolfgang provides the composer with a rich set of options for the motivic development of a given musical theme, allowing the composer to quickly examine and plan multiple strategies while concentrating on a given creative thought. This feature is of great value to composers and producers in advertising, television, records, and film who work under time constraints. To date, the users have found the application useful and have consistently agreed that Wolfgang's compositions communicate the intended emotional quality requested.

Deployment and Benefits

Currently, two composers in advertising and one composer-producer are using Wolfgang in their work environment. The payback and benefits provided by their use of Wolfgang are the following: (1) Wolfgang provides a means to save valuable user-composer time when experimenting with a musical idea. (2) Wolfgang can be used to stimulate creativity. These two application artifacts are exceedingly valuable to a composer.

In a business environment, a composer must continually be creative under varying time constraints. Many composers seldom have the time or freedom to wait for creative thoughts to occur; realistically, their creative output must meet specific project windows. Wolfgang serves as an excellent collaborator-catalyst when a composer's creativity is in a search mode, and time is an important factor.

For example; a composer can use Wolfgang to compose multiple compositions based on a specific motif or theme song of a given customer for an advertising jingle with a given emotion. Wolfgang quickly composes and provides the composer with a combination of different ways to develop or experiment with the motif. Some of these musical examples provided by Wolfgang might even present new musical ideas that the composer might never have considered or might have overlooked, thus stimulating creative thought.

Wolfgang began as a graduate research project in August 1987 and has since evolved into the current system presented in this article. During the first year of work, several prototypes were developed. In May 1988, the current system architecture based on the K-line theory was implemented, replacing a rule-based system architecture. This new design allowed Wolfgang to encode knowledge without having to hard wire new rules into its memory, resulting in a minimal set of explicit rules in an evolving system. From May through August 1988, this architecture was further refined. By September 1988, the application was ready for trial use.

During the design and development of Wolfgang, I served as both the knowledge engineer and domain expert. Prior to my affiliation as a member of the technical staff at Bell Communications and my graduate studies in computing science, I was a professional composer-musician with such credits as undergraduate studies at Juilliard and the Manhattan Schools of Music, world concert tours, and employment as a staff composer for United Artists.

During the first year of prototyping, an average of 15 to 20 hours a week was spent on research, reading, design, and implementation. Because Wolfgang was a personal research project, prototyping was limited to evenings and weekends. I estimate that prototyping under normal working hours would have taken less then five months.

From September 1988 to the present, investigative work with Wolfgang has taken two parallel paths: (1) having Wolfgang review one of its completed compositions with the purpose of searching and planning possible improvements to the completed work and (2) allowing other composers to work with the current application.

Two objectives controlled the deployment of Wolfgang for use by other composers: (1) the determination of how well the application works for others, and (2) what types of output from Wolfgang result from its use and evolution in different environments and cultures. As previously mentioned, all the users have found the application useful and correct in composing works of a given emotion for a specific motif. As for the culture issue, the compositions did reflect the (training) habits of the respective trainer-composers.

The deployment of Wolfgang requires that the user-composer initialize the emotional properties defined in the E-nodes and then interact with learning exercises for the system. Depending on the level of musical maturity required by the composer, the deployment time can range from several days to several weeks. To improve the deployment capability of the application, work over the past six months has focused on improving the user interface to (1) initialize E-nodes, (2) train the system, and (3) run composing sessions. This work also included the porting of code to run on microcomputer hardware. During this time period, the work averaged from 5 to 10 hours a week.

Computers and analog-digital signal-processing hardware have revolutionized both the artistic and industrial functions of the music profession. Over the past 25 years, researchers and musicians have collectively explored the use of computers to generate and compose musical compositions of specific styles; analyze completed compositions based on a well-formed set of conditions; provide intelligent tutoring on varied musical subjects; assist and annotate the creative ideas of a user (for example, a composer or record producer); perform musical works; assist and complement musicians during live performances, and assist in the engineering and production of recorded and sampled musical data. Investigation of the literature leads to many diverse research directions. The proceedings from the First Workshop of Artificial Intelligence and Music at the AAAI National Conference on Artificial Intelligence in St. Paul Minnesota is rich in examples. The annual proceedings of the international computer music conference (ICMC) is also an excellent reference source.

Conclusions

I believe that the signature of a given composer, the semantics of the music, is reflected in the frequent patterns of activated K-lines that occur over a composer's lifetime. These patterns are a reflection of the composer's disposition and environment. The aim of Wolfgang is to allow its disposition (and subemotions) to compose the music. The disposition formulates sets of emotional metagoals that recreate partial mental states (activated K-lines), providing specific emotional qualities to match and satisfy the disposition and its goals.

The work presented in this article is just a beginning for computation based on disposition. Future work will expand the concept of goal formulation with emotional constraints and develop new applications in other domains. Interest has already been expressed to use some of the concepts that were introduced with Wolfgang for the design of new multimedia tools and interfaces. Several disciplines to be considered are animated graphics, scripting for multi-media services, color selection from workstation windowing color maps for hypermedia browsing, storytelling for interactive video, and navigation systems. I believe that Wolfgang demonstrates an innovative approach to goal formulation and the artificial composition of music. Further, Wolfgang provides a working model of computation to study the cognitive aspects of goal formulation for a creative process.

Acknowledgments

The author is indebted to Marvin Minsky and Tod Machover from the Massachusetts Institute of Technology and John Carson from Monmouth College for their helpful interest in, and comments on, this research project. I would also like to thank Robert Allen, Peter Clitherow, Michael Rychener, and Velusamy Sembugamoorthy of Bell Communications for providing helpful suggestions in the preparation of this article.

References

Austin, L. and Delisa, E. 1987. Modelling Processes of Musical Invention. In Proceedings of the 1987 International Computer Music Conference. San Francisco: Computer Music Association.

Baker, M. 1988. An Architecture of an Intelligent Tutoring System for Musical Structure and Interpretation. In Proceedings of the First Workshop of Artificial Intelligence and Music. Menlo Park, Calif.: American Association for Artificial Intelligence.

Balzano, G. J. 1982. The Pitch Set as a Level of Description for Studying Musical Pitch Perception. In *Music, Mind, and Brain: The Neuropsychology of Music*, ed. M. Clynes. New York: Plenum.

Bharucha, J. J. 1988. Neural Net Modeling of Music. In Proceedings of the First Workshop of Artificial Intelligence and Music. Menlo Park, Calif.: American Association for Artificial Intelligence.

Clynes, M., and Milsum, J. H. 1970. *Biomedical Engineering Systems*. New York: McGraw-Hill.

Clynes, M. 1977. *Sentics: The Touch of Emotions.* Garden City, New York: Doubleday

Conklin, D.; and Cleary, J. 1988. Modeling and Generating Music Using Multiple Viewpoints. In Proceedings of the First Workshop of Artificial Intelligence and Music. Menlo Park, Calif.: American Association for Artificial Intelligence.

Deutsch, D. 1982a. Grouping Mechanisms in Music. In *The Psychology of Music*, ed. D. Deutsch. New York: Academic Press.

Deutsch, D. 1982b. The Processing of Pitch Combinations. In *The Psychology of Music*, ed. D. Deutsch. New York: Academic Press

Ebcioglu, K. 1984. An Expert System for Schenkerian Synthesis of Chorales in the Style of J. S. Bach. In Proceedings of the 1984 International Computer Music Conference. San Francisco: Computer Music Association.

Guha, R. V., and Lenat, D. B. 1988. CycLing: Inferencing in Cyc, Technical Report ACA-AI-303-88, Microelectronics and Computer Technology Corporation.

Jackendoff, R. S. 1987. *Consciousness and the Computational Mind.* Cambridge, Mass.: The MIT Press.

Jones, J. A.; Scarborough, D. L.; and Miller, B. O. 1988. An Expert System for Music Perception. In Proceedings of the First Workshop of Artificial Intelligence and Music. Menlo Park, Calif.: American Association for Artificial Intelligence.

Leman, M. 1988. Sequential (Musical) Information Processing with PDP-Networks. In Proceedings of the First Workshop of Artificial Intelligence and Music. Menlo Park, Calif.: American Association for Artificial Intelligence.

Lenat, D. B. 1982. The Role of Heuristics in Learning by Discovery: Three Case Studies. In *Machine Learning*, eds. R. S. Michalski, J. G. Carbonell, and T. Mitchell. Palo Alto, Calif.: Tioga Press.

Lerdahl, F., and Jackendoff, R., A. 1983. *Generative Theory of Tonal Music.* Cambridge, Mass.: MIT Press.

Levitt, D. A. 1985. A Representation for Musical Dialects. Phd. diss. AI Laboratory, Massachusetts Institute of Technology.

Marsella, S. C.; Schmidt, C. F.; and Bresina, J. L. 1988. A Problem Reduction Approach to Automated Composition. In Proceedings of the First Workshop of Artificial Intelligence and Music. Menlo Park, Calif.: American Association for Artificial Intelligence.

Maxwell, H. J. 1988. An Expert System for Harmonic Analysis of Tonal Music. In Proceedings of the First Workshop of Artificial Intelligence and Music. Menlo Park, Calif.: American Association for Artificial Intelligence.

Michalski, R. S. 1982. A Theory and Methodology of Inductive Learning. In *Machine Learning*, eds. R. S. Michalski, J. G. Carbonell, and T. Mitchell. Palo Alto, Calif.: Tioga Press.

Michalski, R. S., and Chilausky, R. L. 1980. Learning by Being Told and Learning from Examples: An Experimental Comparison of the Two Methods of Knowledge Acquisition in the Context of Developing an Expert System for Soybean Disease Diagnosis. *Policy Analysis and Information Systems* 4(2).

Minsky, M. 1979. K-lines: A Theory of Memory, A.I. Memo 516, Artificial Intelligence Laboratory, Massachusetts Institute of Technology.

Minsky, M. 1981. Music, Mind, and Meaning, AI Memo 616, AI Laboratory, Massachusetts Institute of Technology.

Minsky, M. 1986. *The Society of Mind.* New York: Simon and Schuster.

Piston, W. 1941. *Harmony.* New York: W. W. Norton.

Sorisio, L. B. 1987. Design of an Intelligent Tutoring System in Harmony. In Proceedings of the 1987 International Computer Music Conference. San Francisco: Computer Music Association.

Thomas, M. T. 1985. Vivace: A Rule-Based AI System for Composition. In Proceedings of the 1985 International Computer Music Conference. San Francisco: Computer Music Association.

Vercoe, B. 1988. Hearing Polyphonic Music with the Connection Machine. In Proceedings of the First Workshop of Artificial Intelligence and Music. Menlo Park, Calif.: American Association for Artificial Intelligence.

Military

MacPlan: A Mixed Initiative Approach to Airlift Planning
The MITRE Corporation

Naval Battle Management Decision Aiding
U.S. Navy, Space and Naval Warfare Systems Command,
Defense Advanced Research Projects Agency

*C-141 Downloaded by Soviet Worker During
Armenian Earthquake Relief. Courtesy, US Air Force.*

MacPlan: A Mixed-Initiative Approach to Airlift Planning

Kirsten Y. Kissmeyer and Anne M. Tallant

MacPlan is a mixed-initiative, knowledge-based system that helps airlift planners develop resource-effective airlift plans quickly. A mixed-initiative system provides the user with automated help yet allows the user to drive the problem-solving process. The user can select what parts of a plan are to be user developed, and what parts are to be system developed.

MacPlan provides an object-oriented model of airlift entities, a number of heuristic and algorithmic tools for analyzing a plan, and graphic plan-manipulation tools.

Background

The airlift planner's task is made especially difficult by the overwhelming quantities of data involved in the planning process. This information includes data about cargo and passengers, aircraft, operators of aircraft, airfields, and related timing and constraint factors.

A planner's overall task is to allocate and schedule resources to satisfy potentially thousands of movement requirements. A movement requirement is some quantity of passengers or cargo that must be moved from one port to another in a specified time window (figure 1). Three transportation-related categories of cargo exist: bulk, oversize, and outsize. *Bulk cargo* fits on a pallet, *oversize cargo* is too large to fit on a pallet and can only be moved on certain types of aircraft, and *outsize cargo* only fits on a C-5 aircraft.

Each class of aircraft has certain characteristics that affect its ability to support the movement. These characteristics include its capacity (by cargo type and passengers), range, air speed, takeoff and landing requirements, and average allowed utilization rate. Civilian and military operators supply the aircraft apportioned for the movement. Restrictions prevent operators from using certain airfields and, thereby, reduce the number of aircraft that can use the airfield.

Airfields also have a number of characteristics that constrain their use. Cargo and passenger throughput constraints limit the daily amounts of cargo and passengers that can onload or offload at an airfield. Real-world parking and runway characteristics, as well as political considerations, impose additional constraints on airfield use in terms of allowable aircraft types and operators.

The Concept of a Plan

The goal of airlift planning is to develop a plan that provides alternative ways to deliver it's requirements on time. Flexibility is important because airfields can be closed, and aircraft can fail.

We do not use the word plan in the conventional sense of a sequence of actions to perform to reach a goal (Wilkins 1984). An *airlift plan* embodies guidelines for airfield and aircraft use. The following examples illustrate plan guidelines:

- Use airfields A and B for refueling and reassignment of military aircraft after mission completion.
- Disallow civilian aircraft at airfield C.
- Use civilian aircraft for the bulk of the passenger loads, and route them through group 1.
- From the first to the ninth day of the plan, use three 747-100s from operator A, and increase the amount to four from day 10 to the end of the plan.

The movement requirements and the plan, in the form of these guidelines, then serve as the basis for developing a detailed proposed schedule of airlift movements. As described in the next subsection, the current approach to developing such a schedule is to run a discrete-event simulation of the operation.

Airlift Planning Prior to MacPlan

Prior to MacPlan, airlift planning was primarily a manual process. The only electronic aid was a simulation that executed as a batch job on a computer mainframe. The planner developed the plan on paper and

Port of Embarkation:	Airfield A
Port of Debarkation:	Airfield B
Tons Bulk:	30
Tons Oversize:	0
Tons Outsize:	0
Number Passengers:	10
Available to Load Date:	Day 0
Earliest Arrival Date:	Day 1
Latest Arrival Date:	Day 3

Figure 1. A Sample Movement Requirement.

then keyed the information into the computer prior to running the simulation. To develop the plan, the planner reviewed stacks of computer-generated paper reports; plotted onload and offload airfield locations on a map; and drew intended routing strategies on paper charts, incorporating en route stops as necessary to accommodate planes with limited ranges.

The simulation requires that the planner specify routes for each requirement to move from its onload to its offload airfield. To reduce the complexity of the routing network, airfields are grouped geopolitically. In this way, it is only necessary to specify a routing from every requirement's onload airfield group to its offload airfield group. Figure 2 shows the reduction of complexity realized when airfields are grouped in this manner. The dots represent individual airfields, the thin arrows represent requirements that must move, and the ovals represent groupings. The thicker arrows represent the only routes that are necessary to specify routing at the group level. Route distances are then calculated using the longest airfield-to-airfield distance between groups.

The simulation operates upon the resource utilization and routing guidelines set by the planner to produce a detailed schedule of movements. The simulation "loads" cargo onto the aircraft and "moves" the aircraft from airfield to airfield using a set of heuristics to guide the process. The system chronicles any late arrivals or bottlenecks that occur. After the simulation is completed, planners analyze the simulation's results to see how well their specification of resource use moved the cargo and passengers. If something moved late or not at all, the planner must figure out why and what to do about it.

Although the simulation produces accurate predictions on the ability of the plan to execute the operation, it is cumbersome to use for several reasons. No qualitative checking of the plan is performed, and thus, plans that could be identified as logistically infeasible or inconsistent prior to a simulation are simulated anyway. Moreover, the simulation's reports on plan execution tend to be cryptic, making it difficult to determine the source and scope of problems encountered. The combination of the lack of pre-simulation plan-development aids and qualitative checking, as well as difficulty in understanding the simulation's output, necessitated many iterations of modification and simulation before a good plan was developed.

The high turnover rate of expert airlift planners also contributes to lengthening plan-development time. By the time a planner becomes efficient and expert at airlift planning, it is often time to leave for another assignment. For this reason, the Air Force needed a system that would retain some planning expertise and help new planners learn their job quickly.

The MacPlan Concept

When the Air Force asked us to help improve the planning process, it became evident to us that the simulation, although outdated in its technology, still performed its intended function of dynamic analysis quite well. Thus, we chose to focus our attention on the plan-development phase, emphasizing the preparation of a logically consistent and complete plan to be ultimately checked by the simulation. We also realized that for the purposes of deliberate planning, development of the best plan for a given operation was unnecessary. For this reason, we designed MacPlan to be a knowledge-based planning aid that would help produce feasible plans faster than before and could operate as a front end to the simulation process. MacPlan's quality control in plan development reduces the number of lengthy simulation runs to only one or two. MacPlan provides a suite of automated tools for quick plan development and what-if analysis. To further ensure that the plan is ready for detailed simulation, MacPlan contains its own crude internal simulation to detect dynamic difficulties in a plan. Special attention was given to the user interface so that the system could be learned quickly.

Modeling of the Airlift-Planning World

MacPlan bases its model of the airlift-planning domain on the Flavors™ object-oriented extension to Lisp. Flavors can be used to represent an extensive amount of declarative knowledge, and its natural hierarchical organization of classes allows many relationships to be derived by inheritance (Winston and Horn 1989). In addition, the modular construction of Flavors gave us the critical ability to incrementally expand and modify MacPlan. This was particularly important because MacPlan

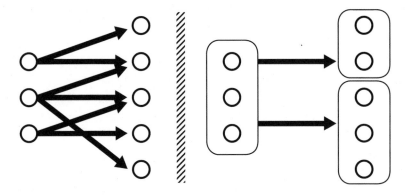

Figure 2. Routing Abstraction.

had to be able to evolve as our understanding of the complex military airlift domain evolved. A metalanguage was also built on top of Flavors and the Lisp substrate to provide constraint-checking capabilities (Abelson and Sussman 1987b).

Objects in MACPlan

The objects in the MACPlan model consist of requirements (cargo and passengers to be transported), aircraft, operators of aircraft, airfields, and routing networks. The plan-element knowledge base is represented as a class-structure hierarchy in which each class of objects can be treated as an object itself. All objects in the model descend from the flavor plan element. The plan-element object infrastructure contains an identifier name and a method for enforcing slot documentation. A plan-element class is created by first defining a flavor of its class name with the desired instance variables. The macro def-plan-element is then called to create the following class structure:

```
(defstruct    (plan-element   (:include
                                    kernel)
                (:type :named-array)
                (:predicate PLAN-ELEMENT?)
                conc-name)
name                        ;from kernel
documentation               ;from kernel
database-mapper             ;map for indexing
                            ;instances in file and memory
instances                   ;list of all existing
```

```
                              ;instances of this object class
    inferiors                 ;all classes that inherit
                              ;attributes from this class
    superiors                 ;all classes that this class
                              ;inherits attributes from
    inst-vars                 ;list of all instance
                              ;variables(slots) local to this
                              ;class and inherited from
                              ;superiors
    ) .
```

All plan-element class objects are either resources or tasks. A *resource* is any entity, such as an airfield or aircraft, that helps to complete a task. A *task* is something that must be done, in this case, something to be moved. They each contain specific key attributes that are used for their respective roles in resource allocation and task execution.

The description of each plan-element class object in the knowledge base consists of three parts: a description of the class of objects to which the object belongs, the flavor instance representing the object to be used for creating instances of the class, and its attributes. Each class description defines its own attributes and inherits any that are defined by ancestors in the hierarchy.

In addition to defining class attributes, each plan-element class description maintains a list of its instances. This list allows the knowledge base to quickly retrieve all instances of a particular class and collect all instances of a class's descendants by traversing the class hierarchy.

An instance of a plan-element class is instantiated or retrieved by calling the following function:

```
(get-object
    plan-element-class-name
    object's-official-name
    attribute-initialization-list
    documentation) .
```

First the knowledge base looks to see if an object of this plan-element class and *official-name* is already instantiated in memory. If not, the knowledge base then looks in its permanent files of information on plan elements. If an object of this class with this *official-name* is found in the permanent knowledge base, the object is instantiated with the specific attributes recorded in its permanent description. Otherwise, the object is unknown to MacPlan, and it receives default attribute values from its class's flavor definition or uses values for attributes as specified in the *attribute-initialization-list*. In addition, get-object fires any initialization methods that have been defined for this class. Initialization methods might be used to create associated display objects or other instances or structures that this class refers to in its slots. Get-object returns two values, the object instance and an indicator of whether it was newly instantiated.

Each attribute, or instance variable, of a plan-element class is defined as an object itself by the macro def-plan-slot. These slot objects are stored in the inst-vars slot of its corresponding plan-element class's structure. By defining slots in this way, we are able to introduce constraints on slots (Huang, Unger, and Fan 1988). Every slot has at least one constraint, a data-type constraint required by def-plan-slot. For instance, a slot can be constrained to contain one of several types of flavor instances, a number within a certain range, or a list comprised of certain types of items. Constraints are discussed in detail in the next subsection.

How MacPlan Reasons

Reasoning mechanisms in MacPlan are closely coupled with MacPlan plan elements (figure 3). Predicates, rules, and constraints form the basic reasoning mechanisms in MacPlan. MacPlan predicates are hand-coded Lisp functions that return two values, either true or false, and a justification for the veracity (a string value). Constraints and rules contain declarative plan-element information. Predicates are only called by constraints and rules but are themselves independent of plan-element representation. Thus, predicates are not affected by changes in plan-element structure.

Forward chaining (Winston 1984) comprises one of the major mechanisms by which MacPlan reasons. The forward chainer uses rules that infer route-planning errors, inadequacies, or conflicts. Rules are a natural medium for describing the event-driven nature of the network routing problem because collections of rules correspond to network conditions (Abelson and Sussman 1987a). The rules are derived from route-planning experts (Hoffman 1987) and describe the relationships between routes and cargo movement. The domain rules are categorized by the cargo type to which they apply and subcategorized by degree of criticality. Rules are true statements such as the following:

• A critical problem exists if there are passengers to be moved along a specific route, and there are no aircraft that can carry passengers.

• A critical problem exists if there are passengers to be offloaded at a specific airfield, but no operator supplying passenger-carrying aircraft is allowed at that airfield.

Such rules are expressed declaratively, as follows:

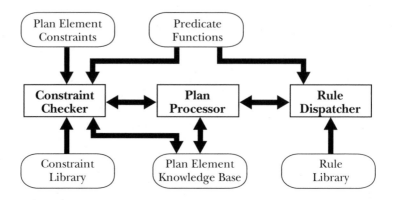

Figure 3. MAcPlan Core Architecture.

(defnetrule critical-pax
   passenger crit-pax critical
   :documentation "Can airfield
   handle at least one passenger
   aircraft and corresponding
   operator?") .

This macro creates a rule object, places it in the appropriate category of similar rules (critical passenger), and places a pointer to the predicate function *crit-pax* in its predicate slot. This particular predicate determines whether the specified airfield can handle passenger aircraft and a legitimate operator.

Rules are applied to the routing network when the user invokes the network analysis tool. The rule dispatcher is activated when the consistency of a route is questioned. It invokes the appropriate category of rules, based on cargo type expressed in the antecedents, and returns the results of the associated predicate function that includes an explanation of any problem encountered. Rules within each category are fired by the rule dispatcher in order of decreasing criticality.

We kept MAcPlan's rule base small and manageable by limiting rules to only represent network routing knowledge. Knowledge about plan-element relationships is instead expressed in constraints. Constraint satisfaction (Charniak and McDermott 1986; Rich 1983) comprises the second major reasoning mechanism in MAcPlan. The airlift-planning process can be viewed as filling slots subject to constraints (Brown 1987). A constraint says that if a certain condition is true, then the contents of certain slots must satisfy certain conditions, and certain other conditions cannot be true. Each slot of a plan-element object is, in it-self, an object containing slots for its value and any constraints on its value. A constraint defines the relationship that holds among a set of slots (Stefik 1981). Constraints can also be maintained between slots of different plan-element classes. A plan-element slot can have many con-straints, which are run automatically when the slot is altered.

Constraints are built as objects; so, different constraints are defined as different classes of objects. Instances of the same type of constraint can thereby be used for different classes of objects in the plan-element hierarchy. The constraint object is pushed onto the constraint slot of the plan element slot(s) whose value(s) it will constrain. Constraints are declaratively expressed as follows:

```
(defconstraint min-max-launch-interval
      requirements
      :min-launch-interval
      :max-launch-interval
      :shorter-eq
      "min-max") .
```

The predicate :shorter-eq determines whether the time interval for the slot :min-launch-interval is longer than the time interval for its related slot :max-launch-interval for any object of class requirements. Enforc-ing constraint checking at the slot level was accomplished by *metalin-guistic abstraction,* or the creation of a language on top of another lan-guage. We built on Lisp's innate message-sending facilities to base constraint checking on plan-element slot access. Each time a message that sets a plan element's slot is sent, the constraint checker is fired and works through the set of constraints identified for this slot. A mes-sage to a slot is built as follows:

```
(SEND plan-element-instance
      slot-message
      [new-value]
      [setting-indicator]
      [justification]) .
```

The setting-indicator determines if constraint checking should be per-formed and how to apply the new value to the slot. The default value indicates constraint checking should be performed, and provided all constraints are satisfied, the new value should be placed in the slot. Other setting-indicator values indicate that constraint checking should be (1) performed but the new value should not be placed in the slot; (2) bypassed with the new value placed in the slot; (3) performed, and provided no constraints are violated, the new value should be deleted from the slot; (4) performed on new value as a member of a set, and

Operation Haylift, 1987. Courtesy, US Air Force.

Operation Haylift, 1987. Courtesy, US Air Force.

Medical Supply Download from C-S in Africa. Courtesy, US Air Force.

Portland Upload for Alaskan Oil Spill. Courtesy, US Air Force.

provided no constraints are violated, the new value should be inserted into the set; and (5) performed on the new value as a member of a set, and provided no constraints are violated, the new value should be excluded from the set. What happens when a particular constraint is violated depends on what type of constraint it is. In some cases, a resumable error, with an explanation of the violation, is signaled. In other cases, the new value is simply not applied to the slot.

MacPlan Features

Plan-Development Aids

MacPlan provides the planner with several features that facilitate fast plan development:

- Movement requirements can be viewed graphically or textually—in the same format they were previously reviewed on paper listings—but with the added ability to select the level of viewing detail.
- MacPlan can group airfields geopolitically to reduce the complexity of the routing network.
- The system can generate direct routes for each movement requirement to travel from its onload group to its offload group. Any legs that cannot be flown by all the aircraft in the plan are displayed in red.
- MacPlan allows the planner to select a region on the map in which the planner would like to use some airfields for stopovers. Any qualified en route airfields in the selected geographic area are brought into the plan.
- A graphic plan-routing tool, called the abstract, provides an interactive electronic version of the previously described method of developing a network on paper charts.
- MacPlan automatically adds all known information about referenced entities to the plan.
- MacPlan tolerates incomplete information. If the system encounters an airfield that it does not know about, it accepts the airfield with default attribute values, allowing the planner to complete the information at leisure.
- Planners can expand and modify MacPlan's permanent knowledge base of aircraft, operators, and airfields.
- MacPlan tracks available versus requested aircraft totals.
- MacPlan provides the ability to save and load plans at any point in development. This ability was accomplished by writing a generic module that can save a list of directly or indirectly referenced objects, ar-

rays, lists, and so on, to file. The file can then be loaded so that the objects in it are reinstantiated to their previous state.

- Planners can download a plan to the external simulation by directing MacPlan to convert its object-oriented plan format to the simulation's record-based format.

Plan Analysis Tools

For quick plan checking, a set of analysis tools was implemented:

- The airlift versus requirements comparison determines the ability of the different types and quantities of available aircraft to move the requirements as a function of time.
- Extraneous network components can be viewed and selectively deleted from the plan.
- A network checker identifies inconsistencies and potential conflicts in the plan. Rules incorporating the planner's knowledge are used to ensure the consistency of the routing network.
- A work load estimator provides a rough estimate of the airfields' abilities to sustain the planned movement.
- An internal simulator generates approximate movement schedules and more precisely identifies the times and locations of bottlenecks in the plan. Although some backtracking is required during the simulation, it is minimized by identifying the repeated invariant constraint checks and performing these checks once prior to scheduling.

User Interface

The user interface was implemented first so that we could effectively evaluate the functionality of the system with the users as it was developed. We made a special effort to provide a natural approach, or interface, to the problem. Thus, the planners played a key role in designing the interface.

MacPlan utilizes a dual-headed, tiled-screen approach to maximize the viewing area and better organize information. Menus pop up in a predictable, context-dependent manner. Color graphics, with structured techniques in the data visualization, portray the plan and results of analysis tools. Layered levels of abstraction and detail are used to represent entities and information in the plan, with the top display level containing the least amount of detail. Clicking the mouse on specific icons reveals further levels of detail. Icons can be directly manipulated to alter the information they represent. Graphs and maps, with extensive color coding, show snapshots of movement tasks versus movement capability at any given time or place in the plan.

The user interface organizes all MacPlan's tools and displays. Natural language, implemented using a semantic grammar, enhances the intuitive nature of the interface. Plan guidelines can easily be entered using natural language, as illustrated by the following:

Disallow C-5s at airfield A.
Recover all military from group 1 to
    group 2 using airfield C
    and airfield D.

Implementation Status

MacPlan runs as a stand-alone, multiprocess, dual-screen system on a Symbolics 3600 Lisp Machine. It is written in Lisp using Symbolics' object-oriented extension Flavors. Two and one-half years elapsed from the conception of MacPlan to its deployment in an operational environment. MacPlan has been operationally evaluated by planners for the past six months.

Evaluation

MacPlan has enabled planners to develop airlift plans an order of magnitude faster than before. Planners often comment that they would have overlooked important plan details if it had not been for MacPlan's help. New planners, even those with little or no experience using computers, come up to speed quickly and find MacPlan extremely easy to use.

Conclusion and Future Work

MacPlan employs an integration of AI, numeric formulas, and graphics technologies to support a mixed-initiative decision aid. MacPlan has successfully enabled new planners to more quickly learn their job.

With the success of MacPlan, our belief in the viability of AI was reinforced. Interest in MacPlan and the problem it addresses identified new fields of research at MITRE in resource allocation and scheduling. Case-based reasoning and problem partitioning, within the discipline of linear programming, are two of these areas. These areas will continue where MacPlan leaves off. Case-based reasoning will perform more of the work for the planner. Although MacPlan allows the planner to reuse a plan for different movement requirements, the planner has to do all the work in identifying an appropriate plan and modifying it to move the new set of requirements. Given a set of requirements, case-based reasoning will be used to select and adapt a suitable existing plan for the planner. This type of reasoning also provides a richly detailed knowledge base for producing explanations about behaviors within the domain.

Although case-based reasoning reduces the amount of planning work necessary by utilizing plans already made, problem partitioning can be used to optimize these plans. MacPlan generates a plan that is feasible but not necessarily optimal. Problem partitioning will be employed to optimize the plan by decomposition.

In addition to identifying further research areas in AI, MacPlan confirmed our beliefs in rapid prototyping with an emphasis on developing the user interface up front. Without an interface to serve as a sounding board, the user typically loses sight of the benefits the system could provide. In addition, the participation of the user is critical throughout the life of the prototype. Reviewing functionality periodically enables the user to see the problem more clearly and suggest additional functionality.

Acknowledgments

We would like to acknowledge the following technical staff who developed MacPlan: Tom Mitchell, Ruben Kleiman, Joe Kartje, Terry Gleason, Tom Doehne. We would also like to thank the following people for their various contributions: project leaders Murray Daniels and Joe Katz; our sponsor ESD at Hanscom Air Force Base in Bedford, Massachusetts; our project officer First Lieutenant Joe Besselman ESD/AV-2; the Air Force planning experts, in particular Major Brian Jones, Major Steve Gordon, and Major Greg Sensiba for their invaluable participation in the project; and General D. Cassidy, commander in chief, United States Transportation Command, and commander in chief, Military Airlift Command, for his support of the project.

References

Abelson, H., and Sussman, G. J. 1987a. Lisp: A Language for Stratified Design, AI Memo 986, Massachusetts Inst. of Technology.

Abelson, H., and Sussman, G. J. 1987b. *Structure and Interpretation of Computer Programs.* Cambridge, Mass.: The MIT Press.

Brown, R. H. 1987. Allocation of Resources in the Knowledge-Based Planning Architecture of CAMPS, Technical Report M87-63, The MITRE Corporation.

Charniak, E. and McDermott, D. 1986. *Introduction to Artificial Intelligence.* Reading, Mass.: Addison-Wesley.

Hoffman, R. R. 1987. The Problem of Extracting Knowledge from the Perspective of Experimental Psychology. *AI Magazine* 8(2): 53-67.

Huang, Y. W.; Unger, E. A.; and Fan, L. T. 1988. Object-Oriented Database Design Using Relational Database Methodology. In Proceedings of the Workshop on Databases in Large AI Systems, 78-91. Menlo Park, Calif.: American Association for Artificial Intelligence.

Rich, E. 1983. *Artificial Intelligence,* New York: McGraw Hill.

Stefik, M. J. 1981. Planning with Constraints. *Artificial Intelligence* 16(2): 111-140.

Wilkins, D. F. 1984. Domain-Independent Planning: Representation and Plan Generation. *Artificial Intelligence* 22: 269-301.

Winston, P. H. and Horn, B. K. P. 1989. *LISP.* Reading, Mass.: Addison-Wesley.

Winston, P. H. 1984. *Artificial Intelligence,* Reading, Mass.: Addison-Wesley.

Knowledge-Based Systems Technology Applied to Naval Battle Management Decision Aiding*

Ted E. Senator, T. M. Witte, and Theodore Kral

The commander-in-chief of the U.S. Pacific Fleet (CINCPACFLT) has responsibility for U.S. naval interests in the Pacific and Indian Ocean regions. He has nearly 300 ships and over 2000 aircraft at his disposal as well as an annual operating budget of over $4 billion. In the management of these resources, he is assisted by a full-time staff of experienced naval personnel who constantly monitor situations, assess capabilities, develop plans, project outcomes, and evaluate strategies. Before assigning resources to missions, they collect, integrate, and evaluate a large body of information from both subordinate units and outside sources and then develop and assess alternative courses of action. Because of time limitations, they normally work toward, and arrive at, satisfactory but not optimal solutions.

---

*The views expressed in this paper are those of the authors and not necessarily those of the Space and Naval Warfare Systems Command, the United States Navy, the Defense Advanced Research Projects Agency, or the U.S. government.

As part of the battle management component of the Defense Advanced Research Project Agency (DARPA) Strategic Computing Initiative, a functional architecture was developed for the application of knowledge-based systems (KBSs) technology to aid and improve this process (Flynn and Senator 1986). An onsite test bed in the CINCPACFLT Command Center was constructed, and AI technology was introduced. Component AI technologies were innovatively engineered into an effective KBS: the force requirements expert system (FRESH). Fresh was integrated with existing operational support systems. It has resulted in more efficient and effective decisions. Its demonstrated utility has led to efforts to transfer its capabilities to other navy command centers. The innovative approach used to develop Fresh has contributed to improvements in navy acquisition policies and procedures for information systems. Other components of the functional architecture are in various stages of development and planning. Challenges remain to add additional decision aid capabilities and transfer them to other command centers with similar but distinct requirements.

Definition of the Problem

CINCPACFLT responsibilities are vast, both geographically and with respect to missions and resources. Mission requirements vary from simple to complex. Some are mutually exclusive, and some are done in conjunction with one another. Some are well defined and others less so. Some are planned years in advance, and some emerge without warning. Resources to match these requirements are units (ships, aircraft, and personnel) and forces (collections of units). Allocation of these resources is a complex problem that is not amenable to linear programming–type solutions. Units of the same type can differ greatly in their designed and actual war-fighting capabilities. The diversity of the nearly 300 ships (aircraft carriers, submarines, destroyers, tankers, minesweepers, and so on) and over 2000 aircraft (fighters, bombers, patrol, land/sea based), as well as the 95 million square miles of ocean under CINCPACFLT's responsibility, emphasizes the magnitude of the problem.

This resource-allocation process consists of constructing forces from units, assigning forces to missions, and scheduling these missions. Decisions can be the result of deliberate planning or crisis response. Employment decisions are based on extensive knowledge about units and missions.

The complexity of the problem and time constraints normally limit a manual search of potential solutions to only the most obvious solutions, and even then, decisions only consider the (ever-changing) most important factors related to the allocation of a resource. This time constraint becomes more severe in crisis situations. This decision problem is made even more difficult by the continual reassignment of experts. Also, this entire process is performed continually in a what-if mode for contingency analysis and planning. Note that in what-if contexts, reasoning over notional forces is required, as distinct from the reasoning about actual forces in the current operations mode.

Brief Description of the Deployed Application

Fresh is a force employment and resource-allocation decision-aiding tool. It operates in both a data-driven and a goal-driven mode. It automatically responds to readiness updates and supports force nominations and analyses of hypothetical new missions and schedule changes. As a resource manager, it monitors the readiness of fleet units, detects significant events, and determines the impact of changes on current and planned assignments. In an analytic role, it generates and evaluates alternative courses of action, including the impacts of potential force redirection. It also acts as an interface aid to less experienced users, providing easier access to real-time operational data regarding readiness, position, and technical characteristics.

In the generation and evaluation of alternatives, Fresh performs a trade-off analysis, weighing knowledge about war-fighting capabilities, mission priority, readiness, planned schedules, operational utilization, personnel endurance, transit times, and fuel cost. The system allows the user to modify these constraints to adapt to changing priorities. It provides explanations consisting of positive and negative impacts of alternatives, including impacts on future schedules of affected units, other ongoing missions, and contingency plans. Alternatives are ranked according to their impacts. Fresh also allows for the nomination of forces to fulfill specified force requirements.

The search space comprehended by Fresh is greater than 10 million elements. It considers the following:

	165	platform/unit types
>	500	scheduled units
	27	employment categories
>	400	employment terms
	15	mission areas
>	100	mission critical equipments
	29	Force level requirements
>	200	command elements
	2540	Pacific ports of call
	120	schedule days/quarter/unit
	365	historical scheduled days/unit
	1825	employment days per unit
>	10^6	ship and unit schedule days
	95×10^6	square miles of ocean

Fresh is an embedded KBS. It is dependent on, and integrated in, a distributed environment with conventional components of the Naval Command and Control System at the CINCPACFLT Command Center. Operational data are provided by the Operations Support Group Prototype (OSGP), a database management, geographic analysis, and display system hosted on a VAX 11/780. OSGP is the gateway for the receipt and storage of both dynamic data (for example, Naval Status of Forces position and readiness data) and static data (for example, ship and port characteristics). These data are automatically replicated on a VAX-11/785 computer system, which acts as a data server for Fresh. The data are organized as an Oracle relational database. Fresh's KBS software is resident on Symbolics workstations, which provide environments for both users and developers. Workstations in the command center itself are tied to large screen displays for group decision support.

The user interface on the Symbolics workstation integrates access to OSGP database, analysis, and display functions as well as Fresh. A menu-based natural language interface based on Texas Instrument's (TI) NLMenu is the primary human-machine interface. It provides 19 interactive screen formats and 13 classes of queries.

Knowledge Base

The Fresh knowledge base comprises interconnected conceptual abstraction hierarchies, constraints, and rules. Alternative generation and evaluation is performed according to a generate-and-test paradigm, using rules for generation and constraints for evaluation.

The hierarchies contain mostly factual knowledge, much of which is derived from information in the relational database. Fresh contains six major hierarchies:

1. Ships, or platform, hierarchy, which contains information about all units of the Pacific Fleet; currently holds 1507 units

2. Activity hierarchy, which contains information about missions and commitments; currently holds 31 activities, or force-level requirements

3. Geography hierarchy, which contains information about ports and bodies of water

4. Command hierarchy, which contains information about CINC-PACFLT subordinate commands

5. Equipment hierarchy, which contains information about equipments and systems; currently holds 427 schema

6. Employment hierarchy, which contains information about scheduling

These hierarchies are interconnected through relations defined by the knowledge engineers. For example, a ship is related to all equipments and systems found on board and is also linked by the operational control relation to the commander of a task group to which it is currently assigned. Inheritance is used within hierarchies.

Constraints represent desirable and undesirable conditions that occur in the knowledge base. Constraints are attached to schemas (objects) or slots (attributes of objects) in the hierarchies. During *impact analysis* (evaluation of an alternative or a significant event), a demon fires whenever a change occurs to the working context of the knowledge base. The demon gathers and evaluates all the relevant constraints (using inheritance within hierarchies and across relations). Fresh currently uses 29 constraints. Rules are used to generate alternatives. Rule processing is initiated by some action, such as the detection of a significant event.

The expert system shell is KnowledgeCraft. (It was selected in early 1985 because of its context facility, which is required for what-if reasoning, and its ability to allow the developer to define the developer's own relations, which was required to interconnect the conceptual abstraction hierarchies.) Prolog is used in the alternative generation functions. Much of Fresh is coded directly in Lisp. Fresh consists of about 176,000 lines of code.

Innovations and Technological Breakthroughs

Fresh's innovation lies in the application and combination of diverse AI technologies. The feasibility of effective interactions between KBS, existing databases, and command and control systems was demonstrated. Technological innovations in knowledge engineering, system engineering, and human-machine interface were achieved.

Knowledge Engineering

Knowledge engineering for Fresh was a formidable job. Knowledge from several domains is required for fleet resource allocation. This knowledge resides in multiple documents and human experts. The manual process that Fresh automates is itself distributed among many people, who are both the experts and the users of the system. Knowledge engineering for Fresh required the integration of knowledge from multiple sources as well as the resolution of conflicts where the multiple sources overlapped or disagreed. Knowledge used for resource allocation required data interpretation by the experts from multiple perspectives prior to incorporation into the system. The prominent breakthrough in knowledge engineering was the use of diverse knowledge representations (that is, hierarchies, constraints, and rules) to handle the different types of knowledge and the creation of the ability to reason across these different data and information types to produce aggregate answers. As the knowledge bases evolved, continued verification was required to maintain consistency and account for the integration of knowledge from multiple sources.

Knowledge and structure had to be incorporated into the system to allow it to deal with the fact that it was a continuously operating (versus transaction oriented) KBS. The system included a mailbox feature for user-defined alerts and continual updating of the current situation knowledge from the database as real world events occurred.

To gain control of the knowledge engineering problem, a methodology of documenting and recording the knowledge in the system (including explanations and sources) was created and instituted. Implementation of this formalism was paramount in achieving success in the effort. It involved creating a knowledge description document to capture and record the knowledge base and a knowledge control board to resolve conflicts. The knowledge description document included the set of facts, constraints, and rules and the relationships among them that are required for the Fresh KBS to reason. The document is a handbook for the validation and verification of the Fresh knowledge base, accessible to both experts and knowledge engineers. This handbook performs the following functions:

It guides experts who have contributed to shaping the knowledge base in examining its contents. It enables communication between the knowledge engineers and the experts regarding the knowledge base structure and contents to facilitate feedback on its accuracy and completeness. Finally, it provides a set of terms useful in discussing knowledge concepts. The knowledge description document was updated approximately every six months. It also facilitated the turnover of knowledge engineers on the project.

In addition, a knowledge review board, consisting of experts and knowledge engineers, and chaired by a senior naval officer, was established. This board had the authority to resolve conflicts between experts.

System Engineering

System engineering of an embedded KBS presented additional challenges that were magnified by the on-site development process. System developers and users-experts worked together in the environment in which the system was being developed and used. Evolutionary development and rapid prototyping were combined with traditional system engineering to develop the integrated system. A large share of the initial effort consisted of the design and construction of the distributed hardware and software environment that supports the KBS components. Fresh consists of four subsystems: KBS, the supporting systems and databases, the human-machine interface, and the communications subsystem. Key areas are discussed in detail in the following subsections.

Evolutionary Development. System development was proceeded by a requirements analysis. Knowledge and procedure-based views of current CINCPACFLT command and control procedures were prepared. Based on these views, a functional architecture consisting of five applications of KBS technology was developed. These notional applications were matched to the available and anticipated technology and the underlying support systems. Fresh was planned as the first of the five applications. The results of this process were recorded in a functional description that served as the basis for Fresh development.

The initial effort primarily consisted of the design of the environment, including the interfaces to the support software. Traditional software engineering methodologies were used, including design reviews and documentation. A functional design document was prepared as a baseline for the design. This document contained the overall system architecture, including hardware, software modules, required data, interface diagrams, and the human-machine interface design as well as the initial design of the KBS components of Fresh. The document has been periodically reissued to keep pace with the evolution of Fresh.

Rapid prototyping, including throwaway demonstrations in conjunction with the design reviews, was used for the KBS components.

A long-term vision guided the overall system architecture, and short-term implementation considerations governed development. A "build a little, test a little, deliver a little" philosophy was adopted to maximize user feedback. Simple productivity-enhancing tools were provided early to demonstrate value to the users and increase their cooperation for the knowledge engineering required for more difficult functions.

Configuration Management. A configuration control board (CCB) was established to monitor the evolutionary development process. Working groups, reporting to CCB, addressed specific areas. The knowledge control board is one working group; others include hardware, databases, and software. System change requests (SCRs) were written by users and developers and prioritized by CCB according to status (error correction, missing required functionality, new functionality), need, and an estimate of required implementation resources. Emergency SCRs could be approved on site for immediate fixes to the current version of the operational system. Software releases occurred approximately each quarter.

Several versions of the system were maintained. An operational version was under user control. A developmental version was used by the knowledge engineers. A tested baseline was maintained for backup purposes. Several demonstration versions, including complete databases and scenario drivers, were frequently used as well.

Dynamic Databases. The original design required KBS to issue calls to the database as necessary during analyses. This approach proved unworkable because of the large number of queries needed to support even a single analysis, communication bottlenecks, contention between processes originating on different workstations, and performance limitations of the database itself. The current design has the data replicated on the workstations at system boot time and included in the knowledge base. Dynamic updates are provided to all workstations as they occur at a typical rate of several thousand per day. Readiness monitoring, which uses these same updates as triggers, is typically turned off on all but one workstation.

Testing. Testing of Fresh was difficult because of the dynamic databases. A valid test of KBS required complete knowledge of the state of the Pacific Fleet. Further, to evaluate the results of a test, a situation similar to one that actually occurred was required. Testing was performed by saving a set of database snapshots and scenario drivers representing situations similar to those encountered in Pacific Fleet operations. These scenarios were also used to support demonstrations.

Human-Machine Interface

The Fresh human-machine interface was crucial to achieving user acceptance. It provides extensive explanations and presents an integrated pseudo–natural language front end to both Fresh and the underlying support systems and databases. The human-machine interface is based on TI's NLMenu.

The human-machine interface was designed as a natural extension of the already existing OSGP interface; this arrangement allowed Fresh's capabilities to be presented as enhancements to an existing system instead of as an entirely new tool requiring additional training. It avoided conceptual distinctions between Fresh functions and existing database, analysis, and display functions. It also allowed an effective and timely exposure of the information contained in the underlying databases because the user did not have to possess an expert understanding of either the multiple database contents and structure or the OSGP command language to extract pertinent information. This format encouraged users to try the Fresh functionality as well. The menu-based natural language interface was natural to use, and it avoided the potential problem—and accompanying user frustration—of asking questions the system could not answer. A natural language integration document defined a Fresh KBS command language for interfaces to the natural language front end, which was analogous to a database interface language (for example, SQL).

Explanations, consisting of the positive and negative impacts of alternatives, not only developed user confidence in Fresh but provided additional utility. In a staff decision-making environment, users carry forward multiple alternatives, along with their positive and negative aspects, to allow for selection by a decision maker. As Fresh was developing, it was able to provide useful analyses even in situations when its absolute evaluations were not yet accepted.

Criteria for a Successfully Deployed Application

Criteria for a successful deployment of Fresh included improved decisions, significant operational use, and a navy program to install it in other command centers. Statistical measures of value added were also generated from comparisons of proficiency (speed and accuracy) between working with and without the system to perform normal tasks.

Initially, Fresh's results were used for comparison with staff member recommendations; now Fresh's results are part of the actual decision making. Fresh has been used to support operations for about two years, both for routine readiness monitoring and user-initiated prob-

Activity	Manual	Fresh
Alternative Rankings	hours – days	minutes – hours
PERSTEMPO	hour	seconds
OPTEMPO	hour	seconds
Readiness Alerts	hours	minutes
Readiness Summaries	hours	minutes
Fuel	hour	second
Deploy Time	minutes	seconds
Transit Time	hour +	minute
Graphic Scheduler		
initially	day	hour
update	day	minute
Special Reports	hours	minute

Table 1. Productivity Improvements

lem solving. The user base has grown to include not only personnel with responsibility for resource allocation but also those with other responsibilities who take advantage of the human-machine interface and the underlying databases. Use has expanded from readiness monitoring and alternative generations to include regular production of eight reports and employment schedules. Fresh is particularly useful in time-critical situations as it systematically considers all possibilities and guarantees that no potential solution or impact is overlooked.

Fresh has matured so that it is commonly accepted and the users look for expanded employment of the KBS technology. Recent use includes mobilization support to CINCPACFLT and his staff during quarterly command postexercises and schedule planning for carriers and battleships for fiscal year 1989. In 1987, the vice chief of naval operations directed that the next-generation operations support system for all navy Fleet Command Centers include Fresh functionality and technology.

Payoff

Fresh is in daily use at CINCPACFLT and accepted into normal operations. The time associated with tasks and decisions has been greatly reduced. By August 1988, improvements in both timeliness and accuracy of decision making had been demonstrated and documented:

- Readiness alerts can be performed in only a few minutes with Fresh versus several hours without it; accuracy increased from 95 percent to over 99 percent.

- Alternatives can now be analyzed and ranked in a few minutes with over 85-percent accuracy as opposed to several hours and less than 60-percent accuracy before. Further, the number of alternatives considered increases from 12 to 100 in a typical case.

- Reports that previously required hours can be produced in 30 seconds with accuracy improved from 90 percent to 99 percent.

- Schedules can be generated in less than two hours with 99-percent accuracy compared to eight hours and 90-percent accuracy without Fresh.

Table 1 depicts typical improvements in decision times provided by Fresh. Speedups in accomplishing tasks range from a factor of 11 for graphic schedule editing to a factor of 400 for alternative ranking. It is worth noting that productivity improvements result not only from the KBS functions but also from supporting functions, such as fuel calculation and report production. A further benefit that was achieved is standardization of the methods for performing some of these supporting calculations.

Direct translation of these improvements into dollar savings is inexact at best. More important than cost savings, however, is the ability to comprehend and make explicit trade-offs between alternatives, resulting in more effective and efficient use of forces. It has been estimated that every 5-percent increase in efficiency results in an improvement equal to $150 million of additional resources, without requiring additional ships, airplanes, or people.

An unexpected payoff arose during Fresh's development and testing. During initial evaluations, Fresh sometimes generated unexpected results. It was determined that KBS was operating correctly but that the data on which KBS based its recommendations was in error. These data had been available as a basis for decision making prior to the introduction of Fresh but were not fully utilized because of the time-critical nature of decision making. Fresh uncovered, and allowed for correction of, these database errors by fully utilizing all available data and explicitly bringing the effects of errors to the users' attention.

System Deployment and Development

Fresh development began offsite in 1985. It was preceded by over one year of architectural development and knowledge engineering by navy operational and technical personnel. Initial onsite installation of test-

Date	Milestone
7/83	Start of preliminary discussions
1/84	Concept initiation
8/84	Formal program agreement
1/85	Letting of development contract
5/85	Preliminary design document and review
9/85	Functional design document and review
2/86	Test-bed equipment installation
6/86	KBS on-site installation
8/86	First on-site demonstration
10/86	On-line operation with real-time data
2/87	First demonstration of operational utility
3/87	Increases in operational user feedback
5/87	CINCPACFLT designated official test bed for decision aids
8/87	Increased functionality causing increases in user feedback
2/88	All development moves onsite
6/88	Memorandum of Understanding for transition
10/88	Prototype completion

Table 2. Fresh Development Milestones.

bed resources and then KBS software occurred in stages during the first half of 1986. A demonstration with a snapshot database took place in August 1986. Fresh began operating with live data in October 1986. Operational utility was initially validated in February 1987. Improvements in availability (50 percent in August 1986, more than 90 percent in August 1987), processing speed, number of alternatives considered, frequency of readiness updates, and number of units have been achieved continuously. Fresh has been used on a regular basis by CINCPACFLT staff members since August 1987. The development effort gradually shifted from the development site to the operational site. Maintenance and enhancement is now performed on site. Table 2 depicts key milestones in the development of Fresh.

Fresh was developed by TI and BTG, Inc. Personnel from navy laboratories and other organizations made significant contributions. Funding and overall direction was provided by DARPA. The space and naval warfare systems command (SPAWAR), DARPA's technical and contracting agent, managed the program.

Direct development cost from 1985 through 1988 was $10 million. These costs do not include the indirect costs of navy experts' time or the costs to develop the underlying databases and support software.

Conclusions

The key to the successful introduction of KBS technology into the navy operational environment was the establishment and use of a test-bed development and evaluation facility in the operational environment itself. This approach allowed for the use of real-time operational data and enabled close coupling between developers and users. The technical management approach for the development of Fresh blended traditional system engineering methodologies with the rapid prototyping techniques that have proven successful in the development of KBS. In retrospect, it is not surprising that a proper blend of these two development methodologies turned out to be successful because the actual system that was developed consists of KBS modules integrated into a larger intelligent system consisting of both conventional and KBS components.

Acknowledgments

As with any large project, success is the result of the efforts of all participants and participating organizations. The foresight of those in DARPA who established the Strategic Computing Initiative and supported the naval battle management application is gratefully acknowledged. We thank the CINCPACFLT staff members, especially Captain Thomas Kenneally, Chuck Deleot, Lieutenant Commander Mike McNeil, and Dr. Raymond Runyan for their assistance, interest, and continued support. The efforts of TI as the primary developer of Fresh are worthy of note. We also thank all the personnel from CINCPACFLT, DARPA, SPAWAR, the Naval Ocean Systems Center, the Naval Electronics Engineering Activity, Pacific, BTG, Inc., BBN Laboratories, Inc., Analytics, Inc., and other organizations for their contributions.

References

1. Flynn, J. P., and Senator, T. E. 1986. DARPA Naval Battle Management Applications. *SIGNAL* 40(10): 59–70.

Operations Management

Expert Operator: Deploying YES/MVSII
*General Motors Corporation and
International Business Machines, Inc.*

Planning the Discharging and
Loading of Container Ships
Information Technology Institute

Expert Operator: Deploying YES/MVS II

R. A. Chekaluk, A. J. Finkel,
E. M. Hufziger, K. R. Milliken, and N. B. Waite.

This article presents the results of research on the use of expert system techniques for automating the operation of large multiple virtual storage (MVS) computer installations.

MVS is IBM's primary operating system for production data processing on large systems. The Yorktown Expert System/MVS Manager (YES/MVS) is an experimental expert system for automating the operation of large MVS computer installations. YES/MVS I, the first version, was used regularly in the computer center at IBM's Watson Research Center (Ennis et al. 1986). YES/MVS II is an experimental collection of components that followed from the lessons of YES/MVS I. YES/MVS II includes support for communication with target computing resources; support for a model reflecting the state of target resources; and an architecture used to structure, modularize, and coordinate automation applications. These components are intended to ease the effort required to implement automation at many different installations.

Expert Operator is an expert system based on YES/MVS II and implemented and placed into production use at General Motors Research (GMR). Expert Operator is the first deployment and test of the combined components in YES/MVS II.

The Challenge of Large System Operations

The challenge of operation and management of large MVS installations and networks has been described at length elsewhere (Milliken et al. 1986; Mathonet, Van Cotthem, and Vanryckeghem 1987). Operating large computing systems and networks is similar to automatically controlling many other processes (Chester, Lamb, and Dhurjati 1984; Nelson 1982; Fagan 1980; Klein, and Finin 1987), but the sources of the unique challenge in operating computing and communication resources are the large scale of such installations and the need for high performance and high reliability.

Because of the numbers of processors, devices, communication links, and software subsystems, the volume and variety of work done, interactions in the work, and requirements for short response times and almost continuous availability, large MVS installations are complex. Console operations play an important role in both performance and reliability in a large installation. Hence, console operations are a challenging problem in real-time control of a large, complex system.

Because of the important role of console operations, the desire is to automate the function to ensure fast and accurate response according to installation policy. In spite of the complexity, it is possible to automate console operations at a particular installation. However, designing one software facility in advance that automates console operations at many sites remains difficult for two reasons: (1) Different sites have different configurations, work loads, and policies. (2) At a given site, the configuration, work load, and policy change with time.

The goal of YES/MVS II is to provide components that reduce the effort required to automate at many installations and maintain and enhance an automatic operator at one site.

Background

In 1982, a research project at the Watson Research Center was started to investigate the use of expert system techniques as a base for automated operations. The result was YES/MVS I (Ennis et al. 1986) developed in OPS5 (Forgy 1981).

YES/MVS I established the feasibility of automating complicated operator actions such as resource management, problem diagnosis, and recovery using rule-based programming techniques. However, it became clear that improved tools and structure could significantly enhance productivity in developing facilities that automated operator actions. Several years ago at GMR, the requirements of MVS console operations had begun to challenge the human operators. The data center had grown to several MVS systems, the individual systems had grown more complex, the systems had significant interactions, and the console message traffic had grown to over 60,000 messages per day.

The first effort to assist human operators was based on a library of short programs in an interpretive procedural language called CList (Command List). The CList programs (working with the network communications control facility (NCCF) and NetView subsystems of MVS) provided significant relief (Chekaluk and Hufziger 1987).

This effort, however, soon reached practical limits. Basically, in the CList environment, a particular console message triggered a particular CList routine from the library. This CList routine executed and handled the message. For responding in fairly simple ways to just one or a few messages, this approach proved reasonable. Difficulty was encountered when the operational task to be performed required multistep actions and knowledge of the current global status of the installation. This experience led to the exploration of expert system techniques and the adoption of YES/MVS II as a base.

Overview of Improvements

The effort to develop Expert Operator using YES/MVS II as a base established the value of several advances beyond those of YES/MVS I:

1. Architecture: The current work produced an architecture for the control portion of a system for automated operations, used this architecture in practice, and demonstrated its value.

2. Model and model manager: The work on both YES/MVS I at IBM and CList's at GMR showed the need for a global model of the status of the target systems. The current research formulated general principles for such models and, equally important, rules to manage them; used these principles in practice; and demonstrated their value.

3. Communications: Special-purpose, data communication support was developed to provide high-level control over message passing between automation and the targets.

4. Expert system shell: Because of the need for improved tools, personnel at IBM's Watson Research Center designed and implemented the Yorktown expert system/language 1 (YES/L1) shell (Cruise et al. 1987). YES/L1 is a data-driven derivative of OPS5 and includes PL/I as a procedural subset. YES/L1 provides the real-time constructs and good performance needed for automated operations. YES/L1 is the prototype of IBM's KnowledgeTool product.

YES/MVS II includes the infrastructure for all these capabilities, but the value of the work is established by the success of Expert Operator. Similar facilities and structure should be applicable to the real-time control of many complex systems because the basic ideas are not dependent upon MVS or the automation of operations.

Of the four advances listed here, this article concentrates on the first two, the architecture and the model and model manager.

Overview of Expert Operator

Expert Operator should be regarded as an embedded expert system because it runs continuously in an MVS address space; acts in real time; and executes through cross-memory communications with NetView, a subsystem of MVS.

Internal Structure

Expert Operator is organized into the following components:

- Communications support for the exchange of status information and control commands with targets.
- A status model that maintains a data summary of the current state of target resources.
- The model manager, a collection of rules and routines that establishes and maintains the (status) model. In Expert Operator, the model manager is the largest functional component measured by code volume.
- Control function, a collection of rules and routines that implement operational policy; organized according to a control architecture based on problems, corrective actions or solutions, and a technique for selecting which solution to attempt in a given situation.
- Support for interaction with human operators; the simplest of the functional components and based on NetView panel services.

Communications

The data sent by target systems are in the form of *messages*. Each message contains data that identify the target system that generated the message and show the time the message was generated. Most messages have a *message identifier* that shows the kind of message. Some messages consist only of a single line. Others consist of multiple lines, and the exact number of lines might not be known in advance.

The operator, human or automatic, sends the target systems *queries* and *commands*. Queries merely request information from the target systems. Commands ask that action be taken which will affect the target system. Both queries and commands cause responses (messages) to be generated and returned by MVS.

Expert Operator includes facilities to receive messages from target systems; recognize, filter, and parse messages; and build and submit queries and commands.

The Model Manager

In YES/MVS I, a collection of rules dedicated to a single type of problem issued its own queries to the target systems (to obtain status data) and issued its own commands to the target systems (to take actions). Experience showed this approach toward target interactions caused problems in two respects. First, the rules and routines of the separate problems largely duplicated the function of interacting with the targets. Second, without a consistent view of target resources, attempts to concurrently address several problems could generate conflicting actions.

Thus YES/MVS II was designed to include a central, consistent model of the status of target systems and a collection of rules to manage and maintain the currency of this model. The model manager in Expert Operator is a deployment of the YES/MVS II model manager but was modified and augmented to maintain a model of the hardware and software configuration at GMR.

Model Characteristics

The contents of a model vary over time with the hardware and software configuration of target resources. However, model changes beyond those driven by configuration changes are much less frequent than are changes in policy-dependent logic for handling problems and allocating resources. A functioning model manager that maintains a broadly applicable model can be used by rules that automate a wide variety of operator activities. With such a model in place, the volume of rules that must be written to solve a problem is typically reduced by at least half.

Data in the model are maintained at the granularity that would normally be of interest to a human operator. For example, the current status of individual devices is maintained, but summaries are maintained for the total number of print lines for small jobs in a printer queue (rather than keeping data on individual, small print jobs). For some resources, MVS volunteers (generates) messages indicating all significant state changes. For other resources, fresh status information must periodically be requested with the period depending on the volatility of the data. Still other information is only collected on demand.

Model Manager Organization

When messages are received, the source is identified, and the time is recorded. If a message is uninteresting, then it is discarded. Interesting messages are parsed, and the status information they contain is used to update the model.

The relation between message identifiers and classes in the model is many to many: A given message might update several classes, and a given class in the model might be updated on the receipt of any one of several types of messages.

All queries and commands to the target are issued through the model manager. Control function interacts only with the model and never directly with the targets.

Each query or command to be issued must be planned in advance and explicitly supported. Every planned query and every command is described in its own PL/I data structure called a *query block*. A query block anchors a linked list of message identifiers of all messages that could be generated by the query.

Any rule or routine in the control part of Expert Operator can request any query. A query can be requested for immediate execution, delayed execution, or periodic execution. A request for periodic execution must be accompanied by a starting time and a period.

Model Integrity. The crucial issues in the model and the model manager are the integrity and currency of data in the model. Currency for a resource is maintained by one of three approaches: (1) frequent, periodic queries, (2) queries on demand before each reference to the data by control code, and (3) the capturing and recording of all state changes as these changes are broadcast in MVS-generated messages.

Achieving integrity involves several considerations, including the following:

If a query (or command) has been issued, but the response messages have not arrived and been processed, then the query is said to be *active*. Because the status model might be in an incoherent state when a response is only partially processed, data in the model that are potentially affected by a query are marked as invalid while the query is active.

For many queries, when the query has been issued, but no response has been received, repeating the query could be harmful. The model manager either serializes such queries, or it recognizes that the response from the first query will suffice for the second query.

The meaning of a response can be ambiguous unless the query is also known, so the model manager does not update the model based on responses that the model did not request. However, some important messages are generated by MVS on the detection of events by MVS. These are recorded in the model whenever they are received.

A query that remains active for an unusually long time can be a symptom of a problem with the target systems. To detect this situation, a rule is written to fire whenever a query has been active too long. The record of active queries must be cleaned up by this rule. (This concern is similar to the situations reported in Fox, Lowenfeld, and Kleinosky 1983).

Control Architecture

Although the model manager is the largest component in Expert Operator, the control component is more likely to change with changes in operational policy, and its organization is critical in simplifying the development of automation applications.

The goals in adopting an architecture for control include the following:

- A modular framework can organize and compartmentalize the development of rules and routines to solve a new operational problem.
- The real-time environment of Expert Operator means that resolutions of several problems are often in progress at once. The architecture should provide a framework so that processing for multiple problems can be interleaved.
- An architecture can organize the resolution of conflicts encountered in the concurrent solution of different problems.
- Partitioning Expert Operator by function into modular components with clean interfaces makes the facility easier to understand, enhance, and maintain.

Solving One Problem

Consider an experienced operator working on just one problem. Some approximation to the following steps would be taken:

1. Detect an exceptional condition in the target systems, that is, a problem.

2. Make initial hypotheses about the problem.

3. Query the target systems for more information if necessary.

4. Revise hypotheses based on new information.

5. Attempt a solution.

6. If this attempt did not solve the problem, return to step 3.

For work on a single problem, we want our architecture to conform essentially to these six steps. In particular, the central part of the architecture is based on *problems* and *solutions*.

Problems and Solutions

Whenever a problem is detected in a target system, Expert Operator creates an instance of class PROBLEM. This instance is deleted when the problem is known to have been solved. The PROBLEM instance serves as an anchor for all other data relevant to solving this problem.

A *solution* is a planned course of action intended to solve a problem. Such a course of action need not merely consist of a command or two to the targets. Instead, it can be of arbitrary complexity. In particular, a course of action can involve examination of data in the model, requests for various queries to update the data in the model, additional inferences from the results of these queries, commands to the targets, more queries to follow up on the results of the commands, and so on.

For each solution, there exists a class of instances. The existence of an instance in a class records the applicability of this solution to an existing problem. The course of action that is a solution is encoded in rules which are enabled by an eligible instance of the solution class. A field on the solution instance records the eligibility of the instance.

A solution instance is always linked to the problem instance for which it records applicability. When a problem is detected and a problem instance is created, the detection process initiates the selection of applicable solutions and the creation of corresponding solution instances. Several solution instances can be associated with a problem instance. Each such solution instance is initially marked as ineligible to solve the problem.

Selecting and Enabling a Solution

Each solution has three ratings: (1) problem severity, that is, the severity of problems for which this solution is appropriate; (2) expected effectiveness of the solution; and (3) impact, that is, a measure of the disruption caused in the target system by applying this solution. The values of these ratings are recorded in the member of the solution class and can be adjusted dynamically.

Because of the variety of possible situations to which one solution could apply and the potential for dynamic change to solution ratings, the choice is made about what solution to apply only after a problem occurs. A particular rule, named the *metarule,* makes the selection and sets the selected solution to execute. From among all solutions instances created but not yet eligible, the metarule selects the solution with the highest problem severity rating, breaks ties by selecting the solution with the highest effectiveness rating, and breaks ties by selecting the solution with the lowest impact rating. Having selected a solution in this manner, the metarule sets the field in the solution instance that indicates that the solution is eligible. Rules that match against this solution are then enabled to fire and take corrective action. This approach is similar to that suggested or adopted by other developers, for example Kastner (1983), Slagle and Hamburger (1985), and Klein (1985).

Interaction with Human Operators

A problem can be in one of two modes, *active* or *advisory.* A selected solution inherits the mode of its problem. In active mode, solutions go ahead and take the actions on the targets as planned. In advisory mode, solutions merely advise the operator about the recommended action to take.

In advisory mode, the actions recommended by a solution are organized into units called suggestions. A *suggestion* is implemented as a linked list of the commands to the targets recommended by the solution. Each recommendation presented to an operator consists of all the commands associated with a solution.

For each problem, solution, or suggestion, Expert Operator can have one line of prewritten, parameterizable English text. An operator explanation facility supports the display of this English text for any instance of a problem, solution, or suggestion. For example, if an operator makes the request

EXPLAIN SUGGESTION 12,

then Expert Operator might respond

THERE ARE TOO MANY JOBS ON THE MVSA SYSTEM
THEREFORE, I AM INVOKING SYSTEM CLEANUP ROUTINES
THE SUGGESTION WILL INVOKE A CLEANUP ROUTINE FOR CLASS T OUTPUT

The text is parameterized by the insertion of short character strings, typically when the instance of the problem is created.

Rationale

This architecture should be evaluated in terms of the motivations for an architecture listed previously.

Compartmentalization. When one starts to attack a new problem, one already knows that problems, solutions, and suggestions will be the major elements, and each of these three will have one-line explanations. Each solution needs three ratings, and for each solution class, there will be a collection of rules which matches against its members. The solution can be debugged in advisory mode. This knowledge is enough to constitute a good running start toward the development of a solution. This same compartmentalization makes activities easier to understand and maintain.

Interleaved Activities. For the rules in the solutions, the right-hand sides are typically short pieces of code. When a rule has completed firing, the KnowledgeTool inference engine selects another rule to fire, and the next rule can be associated with a completely different activity. Thus, the KnowledgeTool inference engine acts as a kind of intelligent task dispatcher. (The solution members in working memory act like task control blocks, and the inference engine's conflict set is something like a task ready list.) Thus, work on several concurrent problems is interleaved.

The Coordination of Independent Solutions. It is also possible to achieve significant coordination of separately developed solutions: For example, suppose suddenly a high-priority problem occurs. Suppose part of the solution for this problem is to restrict other queries and commands to the targets until this problem is solved. Then, one merely writes a rule subroutine to mark lower-priority solutions as ineligible until the high-priority problem is solved.

As a second example, suppose a solution is under way, but suddenly new data indicates that another solution procedure is preferable. Then, it is possible to mark the eligible solution as ineligible and the preferable solution as eligible. Entire problems can be permanently eliminated in midstream when appropriate in light of new data.

Sample Domains

We briefly describe some of the domains of MVS operations currently addressed by Expert Operator.

Backlogged Jobs

One of the major subsystems of MVS is job-entry subsystem (JES). There are two versions, JES2 and JES3. GMR uses JES3, and Expert Operator automates a subset of JES3 operations.

Jobs enter the system from various sources and then work their way through several queues. Each queue is work for a dynamic support program (DSP), a component of JES3. The number of jobs for each DSP can be obtained by means of the JES3 command 8I B. The following is a sample of output from an 8I B issued to a JES3 system:

8I B

IAT8688	0000(w)	0506(A)	OUTSERV
IAT8688	0005(w)	0084(A)	MAIN
IAT8688	0000(w)	0002(A)	WTR
IAT8688	0000(w)	0001(A)	DC
IAT8688	0000(w)	0001(A)	RJP
IAT8688	0000(w)	0003(A)	INTRDR
IAT8688	0000(w)	0009(A)	NJE
IAT8688	0000(w)	0001(A)	NJECONS
IAT8688	0000(w)	0001(A)	TRIG
IAT8688	0000(w)	0001(A)	WATCH
IAT8688	0000(w)	0001(A)	WTP

These messages can be identified by the IAT8688 identifier. Each line of output gives data for one DSP, and the name of DSP is the last token on the line. The numbers show the number of jobs either waiting (W) or active (A).

The number of jobs active or waiting in each DSP and the total number of jobs in JES3 should be kept under thresholds. When numbers exceed thresholds, they are said to be backlogged.

Expert Operator detects a variety of abnormal situations based on the contents of the DSP queues. Considerations include numbers of queued jobs; thresholds; active or waiting; DSP; the total number of jobs in JES3; and so on. Many corrective actions are straightforward, but the variety of possible, abnormal conditions is interesting.

Converter Interpreter Monitoring

One of the most important DSPs is the converter-interpreter (CI) that reads the job control language. CI DSP executes so fast that its queue is usually empty. For example, the command 8I A DCI yields

8I A DCI
IAT8520 NO JOBS ACTIVE ON CI.

However, in unusual situations, for example, a lock is held on a crucial data set, jobs can backlog in the CI queue.

Expert Operator checks the CI queue every three minutes, and it purges and cleans up after any jobs that are in CI longer than one minute.

Disk Reader Monitoring

A *disk reader* (DR) is a special type of job in a JES3 system. Installations commonly write DRs for utility or monitoring tasks. One common technique is to write DR to run, then just before it stops, resubmit itself to run again at some later time. For example, a DR could be written to run once an hour on the half hour but only on weekdays.

Sometimes, a new DR contains logic errors. One common error causes DR to resubmit itself as fast as possible. This error can quickly fill JES3 with unwanted jobs and affect the operation of the entire system. For such a problem, fast response is needed from an operator.

Expert Operator monitors the total number of DRs in the system. If this number becomes abnormally large, then repeating DRs are individually identified and amended by Expert Operator.

Development Effort

The work on YES/MVS I consumed approximately 15 person-years of labor. In comparing complexity of function, it is estimated that Expert Operator delivers approximately 40 percent of the volume of function in YES/MVS I. However, the work at GMR on Expert Operator consumed about 1.5 persons for about 15 months. Of these 15 months, most of the time was spent enabling software outside the central structure of control actions. With the enabling software complete, additional control function continues to be added with much less effort.

Because of the feature of advisory and active modes, the system was used in production in advisory mode almost immediately after rules were written. Production use in full active mode began only a few months after the initial rule development

Summary

Expert Operator is currently in production use. On weekends and other off-shift times, the system runs unattended.

All but the most recently developed problem-solving rules in the system currently run in active mode, providing explanation to operators only as requested. The problem-solving function in Expert Operator was developed by a few skilled people in a few person-months of effort. Function in YES/MVS I to solve problems of similar complexity required markedly greater development effort.

The model manager and control architecture partitioned and organized the system's rules and routines into components by function and the type of external changes that require adaptation in the automation facility. The individual components of the system have proven to be self-contained, easily understood, and comparatively easy to enhance and maintain.

We and others (Strandberg et al. 1985) observed it is often difficult to accurately measure the advantages that accrue to a software assistant. Nevertheless, it is clear based on the function implemented to date that Expert Operator has improved the reliability and availability of the MVS production computer systems at GMR, and the demand for attention from the human operators has been reduced. The project has provided an open path for the future automation of additional aspects of console operations and systems management.

These preliminary results based on one use of the components and architecture of YES/MVS II at GMR indicate that marked improvements have been made in reducing the complexity of function that must be provided by users who want to automate operations at a particular MVS-based computing center.

References

Chekaluk, R., and Hufziger, E. 1987. An Approach to Automated Operation of Complex Computer Systems, Research Publication GMR-5891, General Motors Research Laboratories.

Chester, D.; Lamb, D.; and Dhurjati, P. 1984. Rule-Based Computer Alarm Analysis in Chemical Process Plants. In IEEE Micro-Delcon 1984: Proceedings, the Delaware Bay Computer Conference.

De Jong, K. 1983. Intelligent Control: Integrating AI and Control Theory. Paper presented at the IEEE 1983 Conference on Trends in Applications.

Cruise, A.; Ennis, R.; Finkel, A.; Hellerstein, J.; Klein, D.; Loeb, D.; Masullo, M.; Milliken, K.; Van Woerkom, H.; and Waite, N. 1987. YES/L1: Integrating Rule-Based, Procedural, and Real-Time Programming for Industrial Applications, 134-139. In Proceedings of the Third Conference on Artificial Intelligence Applications. Washington, D.C.: IEEE Computer Society.

Ennis, R.; Griesmer, J.; Hong, S.; Karnaugh, M.; Kastner, J.; Klein, D.; Milliken, K.; Schor, M.; and Van Woerkom, H. 1986. A Continuous Real-Time Expert System for Computer Operations. *IBM Journal of Research and Development* 30(1): 14-28.

Fagan, L. 1980. VM: Representing Time-Dependent Relations in a Medical Setting. Ph.D. Diss., Dept. of Computer Science, Stanford University

Forgy C., 1981. OPS5 User's Manual, CMU-CS-81-135, Dept. of Computer Science, Carnegie-Mellon University

Fox, M.; Lowenfeld, S.; and Kleinosky, P. 1983. Techniques for Sensor-Based Diagnosis. In Proceedings of the International Joint Conference on Artificial Intelligence. Menlo Park, Calif.: International Joint Conferences of Artificial Intelligence, Inc.

Griesmer, J.; Hong, S.; Karnaugh, M.; Kastner, J.; Schor, M.; Ennis, R.; Klein, D.; Milliken, K.; and Van Woerkom, H. 1984. YES/MVS: A Continuous Real Time Expert System, 130-136, In Proceedings of the National Conference on Artificial Intelligence. Menlo Park, Calif.: American Association for Artificial Intelligence.

Kastner, J. 1983. Strategies for Expert Consultation in Therapy Planning. Ph.D. Diss., Rutgers University

Klein, D. 1985. An Expert System Approach to Realtime, Active Management of a Target Resource. MBA/MSE Thesis, Univ. of Pennsylvania.

Klein, D.; and Finin, T. 1987. On Requirements of Active Expert Systems. In Proceedings of AVIGNON-87, Seventh International Conference on Expert Systems and Their Applications.

Maletz, M. 1985. An Architecture for Consideration of Multiple Faults. In Proceedings of the Second Conference on Artificial Intelligence Applications.

Mathonet, R.; Van Cotthem, H.; and Vanryckeghem, L. 1987. Dantes: An Expert System for Real-Time Network Troubleshooting. In Proceedings of the 10th International Joint Conference on Artificial Intelligence, 527-530. Menlo Park, Calif.: International Joint Conferences of Artificial Intelligence, Inc.

Milliken, K.; Cruise, A.; Ennis, R.; Finkel, A.; Hellerstein, J.; Loeb, D.; Klein, D.; Masullo, M.; Van Woerkom, H.; and Waite, N. 1986. YES/MVS and the Automation of Operations for Large Computer Complexes. *IBM Systems Journal* 25,(2): 159-180.

Nelson, W. R. 1982. REACTOR: An Expert System for Diagnosis and Treatment of Nuclear Reactor Accidents. In Proceedings of the National Conference on Artificial Intelligence. Menlo Park, Calif.: American Association for Artificial Intelligence.

Pan, J., 1984. Qualitative Reasoning with Deep-Level Mechanism Models for Diagnosis of Mechanism Failures, 1984. In Proceedings of First Conference on Artificial Intelligence Applications.

Slagle, J., and Hamburger, H. 1985. An Expert System for a Resource Allocation Problem, 994-1004. *Communications of the ACM* 28(9).

Strandberg, C.; Abramovich, I.; Mitchell, D.; and Prill, K. 1985. PAGE-1: A Troubleshooting Aid for Nonimpact Page Printing Systems, 68-74. In Proceedings of the Second Conference on Artificial Intelligence Applications. Washington D.C.: IEEE Computer Society.

Planning the Discharging and Loading of Container Ships: A Knowledge-Based Approach

Tat-Leong Chew, Andrew Gill, and Joo-Hong Lim

The port of Singapore is one of the busiest and most efficient in the world. Fast and efficient ship planning, which is the generation of schedules for container ship discharging and loading, is an important determinant in ensuring the competitiveness of the port. This task is a complex one of generating schedules to meet, as much as possible, and planning guidelines, such as optimizing equipment use while respecting numerous constraints.

This article describes the ship planning system (SPS), a fielded application at the port of Singapore's container terminal that schedules container ship discharging and loading operations. The planning tool includes a comprehensive graphics interface that reduces the work load of the human planner by taking over mundane ship-planning tasks such as tracking multiple plans (or scenarios) for a container ship or performing constraint verification during planning. In addition, it incorporates an automated planner capable of generating its own

Tanjon Pagar Container Terminal, Port of Singapore

schedules, completing partial schedules generated by human planners, and generating partial schedules to a point where human intervention is required.

The need for human intervention stems from the fact that as an operational system, not only must the automated planner perform at near-expert level in terms of speed and quality, it must also be capable in an interactive fashion of referring decisions that it cannot make to a human counterpart.

Human intervention requires that when automated planning fails to complete a schedule, the state of the plan is intelligible to the human planner so that remedial measures can quickly be applied. This situation implies a design approach that allows the automated planner to reason in a similar fashion to human planners with the use of heuristics and reasoning at multiple abstraction levels.

Therefore, we adopted a knowledge-based design approach where the emphasis is placed on the explicit representation of the objects manipulated and on the reasoning used by a human planner in the task. This approach was used successfully in our system.

The Domain

When a container ship arrives at the port, containers have to be discharged and loaded. The chief pieces of equipment necessary are cranes and transtainers. Container operations at the ship are handled by cranes that move along the length of the wharf. During their stay in the port, containers are stacked in a storage area, known as the yard, where they are handled by transtainers. Trucks carry containers between transtainers and cranes (figure 1).

The task of human planners is to generate schedules based on numerous constraints, such as precedences that exist between container operations because of the way containers are stacked on board ship, transtainer clearances, crane clearances, ship trim and stability, special precedence problems resulting from the need to move transit containers from one part of the ship to another, and timing difficulties resulting from containers connecting with other container ships. At the same time, the schedule must comply as much as possible with planning guidelines, such as minimizing the time needed to complete operations, minimizing crane and transtainer movement, and minimizing the breakup of logical groupings of operations.

The process of ship planning involves the following steps: compiling planning information into a stowage summary (figure 2); splitting (or allocating) containers among the cranes, resulting in a split; and se-

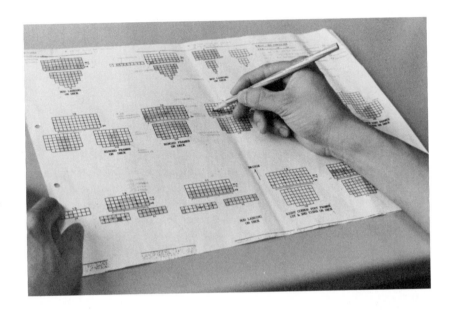

Stowage Profile.

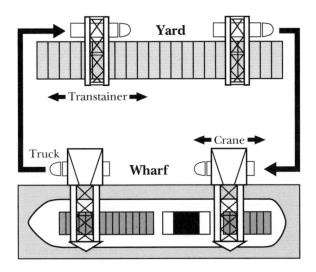

Figure 1. Simplified Schematic of the Yard and a Ship.

quencing (or simulating) the operations while verifying constraints. During splitting and sequencing, any constraint violation encountered is resolved by relaxing either the constraint or another conflicting constraint or guideline. Usually, splitting and sequencing require many attempts before a workable plan is produced.

Main Design Considerations

Based on results acquired from a three-month knowledge-acquisition period with human planners followed by a prototyping phase of another four months, the main tasks identified in the design of a final delivery system were to design an architecture to support interactive planning; develop a representation of physical objects manipulated by human planners during planning, such as cranes and containers; and model and automate the planning process. (Chew, Gill, and Lui 1989).

An Interactive Approach to Automated Planning

As mentioned in the introduction, human intervention is a necessary component in the planning process. One reason is that the human planner is eventually accountable for any compromises in a plan and must be able to guide the automated planner in arbitration between conflicting constraints. In many cases, only a human can accomplish

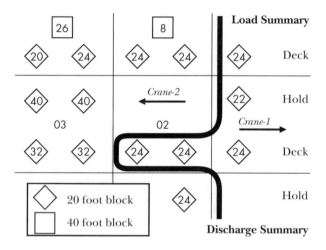

Figure 2. A Simplified Example of a Stowage Summary.
The bold line partitions (or splits) *the blocks*
of containers, between in this case, two cranes.

this role. The planner might have to decide whether a particular constraint violation can be ignored if it is found to severely jeopardize the
quality of the plan.

Another reason is that it is impossible to enumerate ad hoc constraints arising from unforeseen operational considerations that often
occur. Finally, given the complexity of the domain, the initial delivery
system cannot be expected to perform at an expert level at all times. In
fact, a significant amount of refinement by the human planner is expected. The delivery of a black box whose workings are comprehensible only by its designers and, therefore, that requires their intervention
in the event of an unsatisfactory plan would have condemned the planning tool to failure.

Representation for Ship Planning

Object-oriented techniques were used to facilitate the implementation
of an extensive graphic user interface and the representation of the
complex real-world objects manipulated by human planners during
planning. These techniques help manage program complexity through
high-level data abstractions while they provide for a high degree of
code reusability.

Objects represented in SPS include objects that represent the physical structure of the ship, such as bays and superstructures; objects that are used to construct a schedule, such as cranes, containers, and blocks of containers; and constraints that are imposed on objects which have to be verified as they are scheduled, such as precedence or crane clearance.

Our representation of blocks enables the system to model the capability of human planners to schedule large blocks of containers and resort to scheduling with smaller blocks when a constraint violation is to be resolved. This problem is solved in our system by organizing blocks into a tree hierarchy (figure 3).

level of detail

bay, operation type (i.e. whether load or discharge)

— — — — — — — — — — — — — — — —

bay, operation type, row

— — — — — — — — — — — — — — —

bay, operation type, row category (i.e. whether the container is refrigerated, dangerous, etc.)

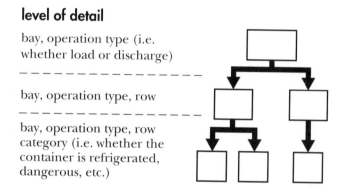

Figure 3. The Block Hierarchy.

Each level of the hierarchy corresponds to a set of container characteristics. Each block at a given level is formed by containers having the same values as these characteristics. Containers at the next lower level are formed by breaking blocks of the level into subblocks with the addition of supplementary container characteristics. The lowest level of blocks (leaf blocks) often consists of one container. All constraints in our system are indexed by these blocks, significantly limiting the constraints considered during verification. Because the leaf blocks of a nonleaf block can potentially be worked in any order, the verification can return violated, not violated, or possibly violated constraints. A possibly violated constraint is an indication to the planner (human or otherwise) that planning ought to be done at a lower level.

Finally, the ease of indexing by data structures that is provided by object-oriented programming techniques is exploited in our system to enable a high degree of maintainability. In effect, a significant amount of the automated planner's code involves the implementation of numerous human planner heuristics for handling constraint violations. Instead of implementing these heuristics in the form of a few monolithic procedures, each handling violations for different types of constraints, the code for the implementation of these heuristics is factored out into manageable chunks indexed by each constraint type. The addition of a new constraint only requires the addition of new code implementing a required set of behaviors (or methods in object-oriented terminology) rather than the modification of existing code. This framework allows quick exploration of the use of the different heuristics in solving constraint violations.

Modeling of the Planning Process

In this subsection, we describe our model of problem solving as observed in human planning. This model is made up of three hierarchical levels: interface, splitter, and sequencer (figure 4). The latter two levels form the automated planner.

Interface. The main screen for the *interface* is an electronic equivalent of the pen-and-paper version previously used by human planners. It allows the user to manually schedule the ship at varying levels of detail from the almost-container level to bigger block sizes. It compiles and presents plan characteristics of all plans explored by the user. At any one time, the interface displays the current plan on which the user can manually schedule operations; any constraint violation is highlighted during this process. In this case, the human planner then overrides or corrects the constraint violation by modifying the existing schedule.

Sequencer. The *sequencer* accepts a split as input and schedules it to discover any constraint violations. It incorporates a series of modules that search for alternative schedules in the event of a constraint violation. Each module incorporates heuristics for localizing the search to specific cranes and blocks. Solutions found (none if no module can be applied), together with the constraint violation, are returned to the splitter, which decides to pursue or abandon the plan.

Splitter. The *splitter* consists of a split generator, which allocates containers to cranes; a diagnostic unit, which handles constraint violations detected when this split is scheduled by the sequencer; and a mechanism for noticing plan failure during automated planning.

The *split generator* implements human planner heuristics in the form of established procedures and practices for allocating containers to cranes based on an analysis of the distribution of containers in a ship. In addition, the split generator receives a set of splitting directives. These directives are used to force the relaxation of planning guidelines during split generation.

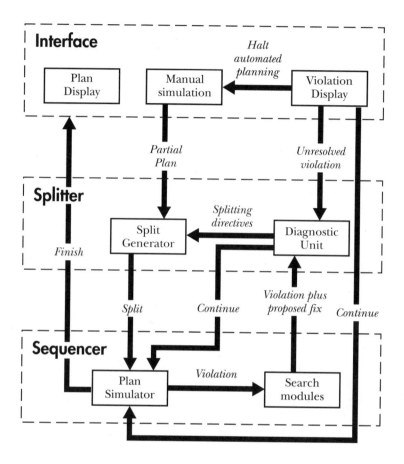

Figure 4. Architecture of the planning tool.

Initially, when a partially scheduled plan is submitted to the splitter by the human planner, no splitting directives, only planning guidelines, are used, resulting in ideal splits in which planning guidelines are closely adhered to. Such splits are rarely workable and cause constraint violations when scheduled by the sequencer. These violations are reported to the diagnostic unit.

The *diagnostic unit* consists of a repertoire of constraint-violation fixing methods. Each method implements a set of heuristics for correcting a specific type of constraint violation, such as crane clearance or tight connection. The output of each method is a set of additional splitting directives to be used in the next cycle of split generation.

Plan failure is noticed by the splitter when no method can be found by the diagnostic unit to resolve a constraint violation or when the same constraint violation is encountered twice by the diagnostic unit. In such cases, the constraint violation is reported to the human planner.

Relation to Other Work

One of the major constraints of the project was to deliver an operational system within two years. Given the short period of time, one of the options was to study the use of AI work in planning, especially domain-independent planners based on Noah (Sacerdoti 1975) and Abstrips (Nilsson 1980) in the hope they could eventually be applied to our problem. However, current research (Swartout 1988; Chapman 1987) indicates that a significant amount of reformulation and reconstruction of such work is still needed. Until this reconstructive work is done, we feel it will be difficult to utilize this work to solve our problem within the time frame.

Our problem-solving approach resembles the approaches taken in the work on Molgen (Stefik 1981a, 1981b) and ISIS (Fox 1986) in that constraints are explicitly represented and manipulated. We found that indexing constraints by the objects they are associated with was a useful technique in limiting the amount of computation. Similarly, indexing the problem-solving procedures by the constraints was most natural. However, we have not formulated problem-solving behavior in our system as a constraint-directed search because human planner heuristics are best represented as procedures.

User Interface

All user interface software in SPS is based on an object-oriented library of graphics routines that simplify code maintenance and enforce interface uniformity throughout the system. This library was developed in Objective-C™, a hybrid, object-oriented C, giving a flavor of Smalltalk-80 to standard C. We also developed a package that elegantly allows the transparent use of these libraries in Lisp-based applications. The package parses the graphic library source files and defines Flavors™ and Flavor methods encapsulating interface code that deals with type-translation details between Objective-C and Lisp and vice versa. Such an approach ensures that the user interface is consistent across all Lisp- and Objective-C based applications; the programmers only need to learn one model of graphic user interface; graphic performance in the

Human Planners at Work With SPS at the Tanjong Pagar Container Terminal

Lisp-based applications is high because the underlying language is C, and the interface layer is negligible in overhead; and maintenance of only one set of libraries is centralized.

Criteria for Success

One of the main criteria that gauges the success of SPS is the system's ability to reduce the current eight-hour closing time. This *closing time* is the cutoff time by which all containers to be loaded on a ship are to be declared to give sufficient planning time to the human planners. To exporters, a reduction in closing time is considered an improvement in the port's services. Much of the closing time in the current procedure is spent in planning the discharging and loading operations and in data transcription between human and computer and vice versa. The aim of SPS is to reduce the planning time by 50 percent and the data transcription to a negligible amount. These reductions will then allow a shorter closing time.

In addition, the volume of container traffic is increasing at a rate that will soon outstrip the capacity of skilled planners using the current methods. A shorter planning time enabled by the use of SPS will allow human planners to cope with the increased work load.

Payoff to Date

The introduction of SPS represents a major change in the established working style and habits of the planners. The system is gradually being phased into their work. Nevertheless, results to date are encouraging. Version 1 of SPS entered operation in June 1988. This version incorporated all the user interface tools but only parts of the automated planner, notably the sequencer and not the splitter. From November 1988 to March 1989 it assisted in the planning of 21-27 percent of the container ships calling at the port. These cases are primarily the simpler instances where the human planners have gained sufficient confidence in the planning tool. In these cases SPS has helped reduce the planning time to less than 50 percent of that previously needed, allowing the planners increased time on complex cases. Almost all the computer-generated schedules were used without amendment in operations. The current plan is to increase the proportion of cases planned with SPS by 10 percent each month and have the system plan progressively more complex cases.

Version 2 of SPS, which incorporates the fully automated planner was delivered for user testing in February 1989 and is now being phased into operational use.

A side benefit of the development effort was the formalizing of the planning methods used by human planners. This formalization now allows the study of these methods for the purpose of improvement.

In addition, SPS ensures a minimum plan quality that the human planner can try to improve on by exploring modifications using the intelligent graphic user interface.

Costs

The project has a budget of US$1.5 million. To date, 19 person-years of development effort have been spent. The project team has grown to 13 members. The team includes an experienced ship planner who has been involved full time since project inception.

Conclusion

SPS is of strategic importance to the port of Singapore in its attempt to maintain its competitiveness and meet the challenge of increasing container shipping volume in the 1990s. The first six months of fielding yielded encouraging results. Not only is SPS becoming a key component in the overall port automation effort, it is also a showcase in illustrating how AI can be applied in an area of strategic economic value to Singapore.

From a technical viewpoint, SPS constitutes an innovative use of interactive hierarchical planning and object-oriented techniques in a complex, real-world problem.

Acknowledgments

We wish to acknowledge the contributions of the SPS project team members and ship planners of the Port of Singapore Authority.

References

Chapman, D. 1987. Planning for Conjunctive Goals. *Artificial Intelligence* 32: 333-377.

Chew, T. L.; Gill, A.; and Lui, E. 1989. Invited paper presented at Sun Technology, Winter 1989.

Fox, M. S. 1986. Observations on the Role of Constraints in Problem Solving. In Proceedings of the Sixth Canadian Conference on Artificial Intelligence, 172-187.

Nilsson, N. J. 1980. *Principles of Artificial Intelligence.* San Mateo, Calif.: Morgan Kaufmann, 350-357.

Sacerdoti, E. D. 1975. A Structure for Plans and Behavior, Technical Note 109, SRI International.

Stefik, M. 1981a. Planning with Constraints (MOLGEN: Part 1). *Artificial Intelligence* 16: 111-140.

Stefik, M. 1981b. Planning and Metaplanning (MOLGEN: Part 2). *Artificial Intelligence* 16: 141-169.

Swartout, W., ed. 1988. DARPA Santa Cruz Workshop on Planning. *AI Magazine* 9(2): 119-130.

Personnel Management

The Ford Motor Company Direct
Labor Management System
The Ford Motor Company and Inference Corporation

The Ford Motor Company Direct Labor Management System

John O'Brien, Henry Brice, Scott Hatfield, Wayne P. Johnson, and Richard Woodhead

The Direct Labor Management System (DLMS) is one major subsystem of a multi-phase Manufacturing Process Planning System (MPPS), which is designed to assist production and planning personnel in all aspects of the manufacturing process; it will eventually support several thousand users. The system is extremely ambitious in scope and has profound implications for the process-planning activity at Ford Motor Company. DLMS is designed to provide the foundation for several intelligent systems aimed at improving the assembly process at Ford and was targeted at several key areas in the planning process. There are five major objectives.

The first is to achieve standardization in process description and improve the clarity of process sheets. A critical document, the *process sheet* is the primary vehicle for conveying assembly information from the initial process planning to assembly at the plant. Process sheets and their derivatives should be an effective means of communicating information at all stages of the assembly process.

Second is to support the creation of work-allocation sheets by automatically generating detailed plant floor assembly instructions from the abstract process sheet. The *allocation sheet* contains the detailed assembly instructions assigned to an individual assembly worker (several allocation sheets are typically derived from one process sheet by elaborating and reordering operations specified on the process sheet).

Third is to automatically provide consistent and accurate estimates of product and nonproduct labor times involved in the assembly process. (*Product times* represent effort directly associated with product assembly. Removing this essentially clerical task frees the industrial engineer to effectively use personal expertise in analyzing the assembly processes. Consistent process descriptions help by highlighting any process inefficiencies.)

Fourth is to provide a foundation for generic processing. Process sheets are currently written at a fine level of granularity. The objective of generic processing is to provide process descriptions at a higher level and, thus, promote standardization in the assembly process across vehicle types. Process sheets can now be written at any level of abstraction. Instructions can be written in terms of the micromotions made by the assembly worker (for example, grasp screw, or position screw) or, alternatively, as macrodescriptions applied to complete vehicle subsystems (for example, install brake system).

Fifth is to provide a foundation for machine translation. The production of automobiles is increasingly becoming an international effort, and the overhead involved in translating process instructions into different languages is considerable. The standard language is designed to facilitate this translation and the taxonomy is structured such that it will serve as an interlingual form for the target languages.

System Architecture

The first stage in system development was to identify a grammar for describing assembly processes. This grammar is outlined in the following subsection. The software system implemented to interpret process descriptions is described in subsequent subsections. The software architecture is illustrated in figure 1.

The Process-Description Language

A clear prerequisite for system development was the identification of a standard format for writing process descriptions. These descriptions were originally written in free-form English and uninterpretable by computer.

Process sheets in the re-engineered system are described using a case grammar developed specifically to meet the requirements of automotive assembly. It provides for the expression of imperative English as-

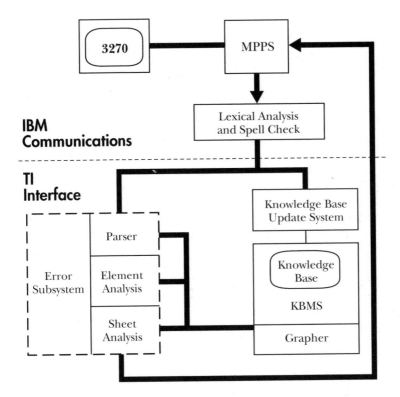

Figure 1. The DLMS Software Architecture.

sembly instructions at any level of detail. The fact that the language had to be accessible to potentially thousands of engineers with different educational levels meant that the necessary coverage had to be achieved without significant loss of expression or naturalness. Approximately 300 engineers have been trained to date to use the language without any significant difficulties. A typical process sheet is illustrated in figure 2.

Process data fall into two categories. Process descriptions are specified in the main body of the text and in part and tools areas of the sheet. These data are supplemented by contextual data; the sheet title and vehicle and plant information, for example.

Certain word categories in the language possess specific semantics; these words were defined by the engineering community. Verbs in the language are associated with specific assembly actions and are modified by significant adverbs where appropriate. For example, the

SHEET EJ540000000 CONNECT HEATER HOSES TO A/C EVAPORATOR ASSY.
10 OBTAIN LUBRICANT FROM STORED POSITION
20 SPRAY LUBRICANT ONTO A/C EVAPORATOR ASSEMBLY NIPPLES
30 OBTAIN 2 CLAMPS FROM STOCK
40 POSITION 1 CLAMP ONTO OUTLET HOSE
50 SEAT HOSE TO A/C ASSY.
60 POSITION EXISTING CLAMP ON HOSE BETWEEN A/C
     EVAPORATOR ASSY. NIPPLE AND ASSY.
70 SECURE OUTLET HOSE TO HEATER NIPPE WITH 1 HOSE CLAMP
80 POSITION 1 CLAMP ONTO INLET HOSE
90 SEAT HOSE TO A/C EVAPORATOR ASSY.
100 POSITION EXISTING CLAMP ON INLET HOSE BETWEEN A/C
     EVAPORATOR ASSY. NIPPLE AND ASSY.
110 SECURE INLET HOSE TO HEATER NIPPLE WITH 1 HOSE CLAMP
     USEING A NUTRUNNER
TOOL (70 110) 1 P AAPTTCA TSEQ RT ANGLE NUTRUNNER TORQUE
SHUT OFF
TOOL (70 110) 1 S 2H12E10H16B56 TSEQ 3/8 HEXSRT BDY1/2
ODLONG LNGTH.
PROCESS-PREFIX EJ
REPEAT-OPPOSITE-SIDE N
PLANT AGO
VEHICLE CQA

Figure 2. A Typical Process Sheet.

fragments *inspect, visually inspect,* and *manually verify* all have different interpretations. Important cases in the language are denoted by specific tokens. Sentence 110 from the example sheet is illustrated with its constituent cases in figure 3.

Other areas of the process sheet containing part, tool, vehicle, and other related information are parsed to provide extra detail and context for the sheet.

The Knowledge Base and Associated Management System

At the center of the DLMS system is a large taxonomy of automotive assembly expertise. This taxonomy maintains descriptions of any concepts required to interpret the surface language and generate the implied atomic work instructions. Concepts in the taxonomy include parts, process equipment, standard operations, and geometric workstation models. A portion of the current taxonomy is illustrated in figure 4.

The DLMS knowledge representation system is implemented on the NIKL (Moser 1983; Kaczmarek, Bates, and Robins 1986) model. All con-

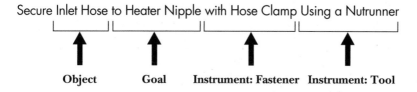

Figure 3. Example Sentence and its Constituent Cases.

cepts in the environment are described in terms of a frame-based description language (FDL): this language is a variation of the FL- language described by Brachman and Levesque (1984). It meets all the identified requirements for terminological reasoning in this application domain and avoids the computational limitations implied by FL. (See Brachman and Levesque 1984 and Nebel 1988 for formal treatments of issues involved in selecting a language to support terminological reasoning).

Concepts in the taxonomy fall into two basic categories: primitive and nonprimitive concepts. *Primitive concepts* (corresponding to natural kinds) are concepts whose structure contains necessary but insufficient criteria for determining subsumption. Thus the system never automatically infers that a supplied concept is subsumed by a primitive concept; it must be specifically instructed to make the connection. The operation concept in figure 4 is an example of a primitive concept.

Nonprimitive concepts are assumed to be fully defined by their descriptions, which contain necessary and sufficient criteria for determining subsumption. Nonprimitive concepts are defined by their roles, which describe properties of a concept by relating it to other concepts. Roles approximately correspond to slots in a frame and form a subtaxonomy in their own right. A role is attached to a domain concept (the most general concept for which the role has meaning) and represents a set of fillers for each instance of this concept. Role restrictions are used to constrain the set of values that a filler can take. The "Secure threaded fastener using power tool" concept from figure 4 is a nonprimitive concept. A partial description of the concept and some of its superconcepts are shown in figure 5 (the notation follows the example of Moser 1983).

Concepts can also be attributed with nondefinitional information. One use of this datatype is to associate a process script with each standard operation.

An important result of the precisely specified semantics of the concept description language is that a formal definition of subsumption and an associated classification procedure can be defined. The classification process involves determining the most specific "subsumers" and most general "subsumees" of the current concept with the taxonomy.

The notion of subsumption is intuitively clear (Nebel 1988):

"Concept 1 subsumes Concept 2
<=> all objects which are a Concept 2 are also a Concept 1" .

The test for subsumption used in this application is as follows:

Concept 1 subsumes Concept 2 if:

• All primitive concepts that subsume Concept 1 also subsume Concept 2.

• For each roleset of Concept 1 and the corresponding roleset of Concept 2, the value description of Concept 1's roleset subsumes that of the value description of Concept 2's roleset.

More detailed presentations of classification in the KL-One family of systems are given in Schmolze and Lipkis 1983 and Lipkis 1981.

The DLMS classifier is incremental in nature. As concepts are added or modified, only those inferences that are required to position the current concept and reclassify any related concepts are made.

Two interfaces to KBMS are provided. A procedural interface was developed to support reasoning within the process sheet and end-user knowledge base update (KBU) subsystems. An interactive graphic interface inspired by the ISI grapher (Robins 1986) supports taxonomy maintenance for system development purposes.

The number of concepts and the level of detail at which they must be represented is determined by the requirements of any standard operations and the need to discriminate between them. The set of defined operations determines the granularity of the resulting plan and, hence, the resolution of any associated time estimates. The required granularity or resolution can only be determined with respect to the processing requirements of any downstream client application. The resolution of the system increases over time as significant derivatives of existing operations are identified. This feature leads to a definition of operational redundancy: a concept is redundant (for the purpose of creating work instructions) if no associated standard operation exists that refers uniquely to it. To clarify this concept, consider the taxonomy fragment shown in figure 5. The concept nut is clearly redundant because the only operation of significance refers to the concept threaded fastener, which subsumes the concept nut.

The Parsing Subsystem

The *parser* produces a structure for grammatically correct sentences and provides information about the state of the parse to the error subsystem when parsing fails. It was primarily designed to support efficient

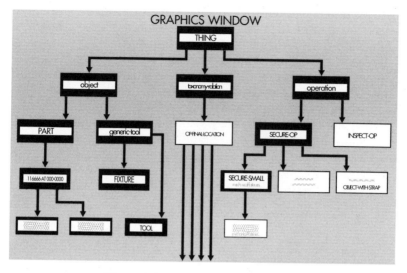

Figure 4. Part of the DLMS Taxonomy.

processing of syntactically correct sentences and provide for flexible extension of the grammar. Because the system is transaction oriented, and no direct user interaction is possible, any errors are processed as a group at a later stage in the analysis.

The parser is implemented as an augmented transition network (ATN) (Winograd 1983; Allen 1987). To improve the ATN performance, the implementation includes a cut operator that eliminates backtracking in cases where traversing a specific arc clearly excludes all other alternatives. In addition, to eliminate the cost of undoing actions during backtracking, the parser defers most of its actions to a second phase, which it performs once a successful parse is achieved. This second phase generates a predicate-based representation of the parse tree used in the element-analysis subsystem described later.

The parser has an associated rule-based error-processing component specifically tailored to match the standard grammar. It detects errors and suggests appropriate corrections to the user. No attempt was made to implement a general mechanism, for instance, of the type proposed by Weischedel and Sondheimer (1983). Error-system requirements were derived by exhaustively analyzing the types of errors made by process engineers during their initial attempts to use the process-description language.

Data are passed to the error subsystem by the parser, which maintains a set of states representing the progress it has made in interpreting a given sentence. This set contains those partial parses having the greatest number of edges. If the parser is blocked, it reconstructs the partial parse tree for each of these states and applies rules to select one as the most likely intended interpretation.

The error system proceeds by using information from the lexicon to identify case markers in the unparsed portion of the sentence. These case markers are then used to hypothesize any constituent phrases intended by the user from clues provided by the marker itself, the partial parse, and the position of the phrase relative to other datum tokens (any verb or direct object for instance). These hypotheses are tested using the appropriate transition subnetwork. When a complete hypothesis for the intended sentence structure is determined and validated, it is translated into an error message and an appropriate correction strategy for return to the user.

The Element-Analysis Subsystem

The objective of element analysis is to generate a set of atomic work elements representing the direct-labor content implied by the sheet. This set can be interpreted as a process script for performing the assembly task. *Direct labor* encompasses all those actions that directly contribute to the product assembly process. *Indirect labor* constitutes all the activities that are required to effect the associated direct actions.

The process sheet illustrated in figure 2 is used to motivate the discussion.

A preliminary task involves identifying any significant contextual data (vehicle, plant, and functional subsystem of the vehicle, for example) from the appropriate regions on the sheet and establishing plausible defaults. The system continues by interpreting the parse trees supplied by the parsing subsystem. The first phase resolves any weak or implied references in the sheet (ellipsis, intraanaphora and interanaphora, and so on).

The next phase is to map individual statements on the sheet into the corresponding operation concept within the taxonomy. Consider statement 110 from figure 2. Objects implied by the constituent noun phrases in the statement are identified:

- Appropriate prepositional phrases are transformed into a normal noun-noun modifier form.
- Noun-noun modifiers are analyzed and individual concepts identified using a precompiled dictionary structure.
- Concept descriptions are created and classified to determine the corresponding persistent taxonomic concept.

Significant cases in the statement are identified from the appropriate tokens and mapped onto their corresponding roles. Any tools associated with the statement are identified as necessary instrumental cases.

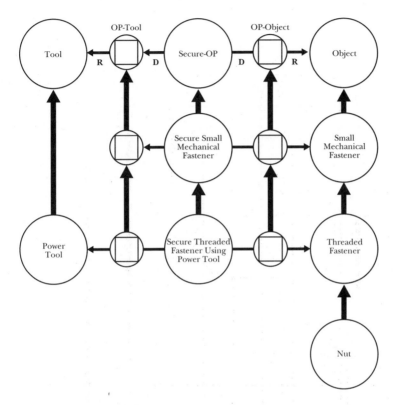

Figure 5. Taxonomy Structure for Secure Operations.

An operation description is constructed and classified to identify its counterpart standard operation, and the assembly script stored with the definition of this operation is retrieved. The operation description created for statement 110 is illustrated in figure 6.

After any necessary variable bindings are established for the script, its constituent activities are, in turn, classified to identify their standard counterparts. This process continues until a set of terminal scripts is identified. This set represents the direct-labor content of the statement. This set is then supplemented by adding any tool-handling information required to undertake the operation.

The set of direct allocatable elements generated by statement 110 is illustrated in figure 7. Redundant data are embedded in the output to maintain contextual information when the elements are split into operator work allocations. The codes in parentheses represent the direct-labor times associated with each element.

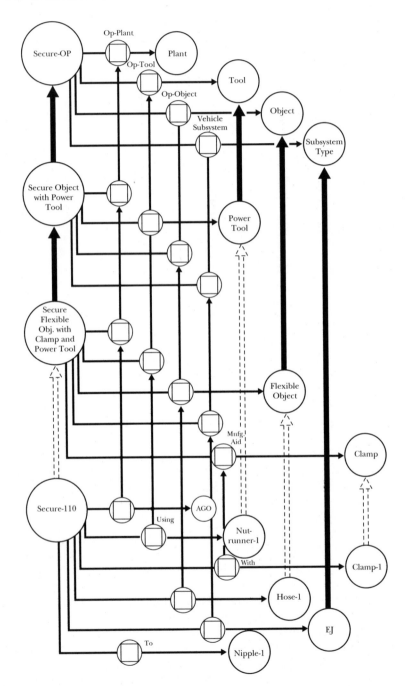

Figure 6. Operation Structure for Sentence 110.

"SECURE INLET HOSE TO HEATER NIPPLE WITH 1 HOSE CLAMP USING A NUTRUNNER."

GRASP POWER TOOL (RT ANGLE NUTRUNNER TORQUE SHUT OFF) (O1M4G1)
POSITION POWER TOOL (RT ANGLE NUTRUNNER TORQUE SHUT OFF) TO HOSE CLAMP
SECURE INLET HOSE (O1M4P2) TO HEATER NIPPLE USING POWER TOOL (RT ANGLE NUTRUNNER TORQUE SHUT OFF) WITH HOSE CLAMP (O1M2P5M1P0)
RELEASE POWER TOOL (RT ANGLE NUTRUNNER TORQUE SHUT OFF) (O1M4P0)

Figure 7. Direct Elements for the Sentence.

The Assembly Process Simulation Subsystem

The *assembly simulator* elaborates the set of direct elements by generating any additional indirect (nonproduct) work elements required to implement the current plan. These elements are essentially the motions the operator needs to make to move around in the workstation.

The first stage in this process is to identify the standard workstation configuration in which the assembly operation is assumed to take place from a taxonomy of standard configurations. This is done by creating a workstation description from contextual information in the sheet and classifying it. Each work element is then processed in sequence to identify any datum locations referred to in the text. The operator is assumed to relocate between datum locations in carrying out the assembly operation. The absence of a datum location for a given element means that the operator remains in situ for this element. Datum locations are once again identified by creating concept descriptions and classifying them. This allows a location to be associated with a concept at any level of detail.

The complete set of elements for sentence 110 of figure 1 is illustrated in figure 8.

Updating the Knowledge Base

The KBU tool is designed to be accessible to any of the process or industrial engineers who use the system. The design goal was to allow the user unrestricted access to maintain that part of the taxonomy, which directly relates to the user's area of assembly expertise. This access was necessitated by the fact that no single source of assembly expertise exists; in fact, several hundred experts in the assembly community contribute parochial expertise to the final objective.

WALK TO POWER TOOL
GRASP POWER TOOL (RT ANGLE NUTRUNNER TORQUE SHUT OFF)
(01M4G1)
WALK TO HEATER
POSITION POWER TOOL (RT ANGLE NUTRUNNER TORQUE SHUT OFF)
TO HOSE CLAMP
SECURE INLET HOSE TO HEATER NIPPLE UING POWER TOOL (RT
ANGLE NUTURNNER TORQUE SHUT OFF) WITH HOSE CLAMP
(01M2P5M4P5M1P0)
RELEASE POWER TOOL (RT ANGLE NUTRUNNER TORQUE SHUT OFF)
(01M4P0)

Figure 8. The Complete Set of Elements for Sentence 100.

KBU is divided into two basic areas of functionality: process and industrial engineering. The *process engineering* community maintains a common dictionary and adds descriptions of parts, tools, and other assembly equipment to the taxonomy. The *industrial engineering* groups are responsible for maintaining the part of the taxonomy that contains standard operations. A query facility is universally available to facilitate knowledge base interrogation.

The Deployment Architecture

The DLMS system is installed within the existing Ford Motor Company systems network. An overview of the back-end processor architecture is shown in figure 9.

The hardware platform is the Texas Instruments (TI) Multi-Processor (MP). MP utilizes a 16-slot Nubus architecture and allows four Explorer II processors to share the same chassis and console. System software supports interprocessor communication and individual processor rebooting. It also allows processors to share a common load band. Communication with the IBM host is effected by way of the TI SNA II processor, which acts both as a 3274 terminal controller and a terminal emulator.

Software Tools

The rule-based system components were implemented in Art,™ from Inference Corporation. All other software was written in Common Lisp.

Current Status and Future Development

The system was developed with continuous input from the process and industrial engineering communities. Because the potential conse-

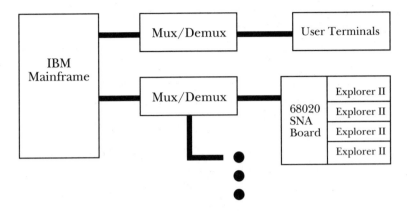

Figure 9. DLMS Deployment Architecture.

quences for the organization are substantial, the system is being deployed incrementally. This approach is critical because the quantity of knowledge required to competently handle all Ford vehicles is huge, and the integrity of the knowledge base is of paramount importance. The system was introduced experimentally in December 1988 to a small subset of the process and industrial engineering community and is currently being used by approximately 300 process and industrial engineers. The system is part of a major initiative at Ford Motor Company and will continue to be added to for a number of years.

Acknowledgments

The authors would like to thank Thomas S. Kaczmarek for his invaluable contribution to the initial specification and design of the DLMS system. Thanks are also owed to Philip Klahr and Philip C. Jackson, Jr., who were extremely helpful in criticizing early drafts of this article.

References

Allen, J. 1987. *Natural Language Understanding*. Menlo Park, Calif.: Benjamin Cummings.

Brachman, R. J. and Levesque, H. J. 1984. The Tractability of Subsumption in Frame-Based Description Languages. In the Proceedings of the National Conference on Artificial Intelligence, 34-37. Menlo Park, Calif.: American Association for Artificial Intelligence.

Kaczmarek, T. S.; Bates, R.; and Robins, G. 1986. Recent Developments in NIKL. In Proceedings of the National Conference on Artificial Intelligence, 978–987. Menlo Park, California: American Association for Artificial Intelligence.

Lipkis, T. 1981. A KL-ONE Classifier, Consul Note No. 5, USC/Information Sciences Institute.

Mark, W. 1981. Representation and Inference in the Consul System. In the Proceedings of the Seventh International Joint Conference on Artificial Intelligence. Menlo Park, California: International Joint Conferences on Artificial Intelligence, Inc.

Moser, M. G. 1983. An Overview of NIKL, the New Implementation of KL-ONE, Rep. 5421, Bolt, Beranek and Newman, Inc.

Nebel, B. 1988. Computational Complexity of Terminological Reasoning in BACK. *Artificial Intelligence* 34: 371-383.

Robins, G. 1986. The ISI Grapher and the ISI NIKL Browser, Internal Memo, USC/Information Sciences Institute.

Schmolze, J. G. and Lipkis, T. A., Classification in the KL-ONE Knowledge Representation System. In the Proceedings of the International Joint Conference on Artificial Intelligence. Menlo Park, Calif.: International Joint Conferences on Artificial Intelligence, Inc.

Weischedel, R. M., and Sondheimer, N. K. 1983. Meta-Rules as a Basis for Processing Ill-Formed Input. *American Journal of Computational Linguistics* 9(3–4): 161-177.

Winograd, T. 1983. *Language as a Cognitive Process*, 195–267. Reading, Mass.: Addison Wesley.

Retail Packaging

Packaging Advisor: An Expert System
for Rigid Plastic Food Package Design
The Du Pont Company

*Packaging Advisor™ Assists the Designer of Rigid Plastic Food
Containers, Such as the Ketchup Bottle this Youngster is Holding.*

Packaging Advisor™: An Expert System for Rigid Plastic Food Package Design

Alvin S. Topolski and Douglas K. Reece

Packaging Advisor is an expert system that designs rigid plastic food containers. It was developed at the Du Pont company to help both our customers and our own staff better understand the use and benefits of our barrier resin products in food containers. The barrier resins business was a new business for Du Pont, and we had new and unique products to introduce to the marketplace. Packaging Advisor was the keystone of the marketing communications strategy for these new products and was recognized by management for a substantial contribution to the success of the new business venture. In this article, we describe the business and technical environment in which the system was constructed, review the system and its development process, and describe how the system was successfully used to achieve our business objectives.

The Business Environment

Rigid plastic food containers offer many advantages to the consumer. They are lightweight, dentproof, rustproof, and shatter resistant and can be molded with such convenience features as handles and pouring spouts. These advantages have led to the extremely rapid growth of plastic containers as replacements for metal and glass.

The size and growth rate of the plastic food packaging market was, therefore, an attractive market for the Du Pont company. Among our offerings to this market are several barrier resin products, introduced in 1987 and 1988. These plastic resins reduce the infiltration of oxygen, which can cause degradation or spoilage of package contents.

The products introduced in 1987 met with limited success for a number of reasons. We were competing with several established suppliers with similar products. The process that a package fabricator uses to qualify a new supplier is arduous: Substantial design work must be followed by trial runs with the new material. Furthermore, as discussed later, the properties of barrier resins are complex, and many details of the application must be analyzed to predict performance.

Du Pont's business plan called for the 1988 introduction of new and technically superior materials that were to be competitive. However, means were needed to establish Du Pont as a credible and technically sophisticated supplier of barrier products and communicate the value of the new products.

To address these needs, the business team developed a concept for a computer system that would automate the process of designing a food package. The system would offer alternative designs using Du Pont's and competitive resins and showing corresponding costs. The system would clearly show the value of the new products, and its ability to automate tedious design calculations would make it appealing to customers.

The Du Pont company has an aggressive program to implement AI applications. Staff members from the AI group agreed to provide the needed system resources, and work was begun in May 1987. The completed system was fielded nine months later in February 1988.

The Food Package Design Problem

Food packages requiring an oxygen barrier are typically manufactured as multilayered structures. The bulk of the package is composed of a structural material selected for its durability and low cost. One or more barrier layers are used to achieve the required limits on oxygen infiltration, and layers of adhesive material are used as necessary to prevent the package from delaminating. The design problem addressed by Packaging Advisor is to select appropriate barrier and structural materials for a given application, determine how much of the (usually more costly) barrier material is required to achieve a specified limit on total oxygen permeation during the shelf life of the package, and calculate the materials cost for the package. Selection of adhesive layers was left to a separate expert system.

The oxygen barrier properties of a material are measured by the rate at which oxygen infiltrates across a unit area and thickness of the material. This oxygen-permeation parameter is a function of temperature and, for some materials, humidity. The humidity to which the material is exposed depends on the humidity inside and outside the package and on the water vapor transmission properties of other layers within the package. Some structural materials provide a significant amount of resistance to oxygen permeation. Thus, the requirements for the oxygen barrier material depend strongly on environmental factors and the other materials with which it is used.

Several other factors must be considered. Packaging materials must be compatible with the intended fabrication process and must have whatever degree of optical clarity is required for the application. Because the layers of the package are extruded together, they all must have similar processing temperature ranges. Federal regulations restrict or prohibit the use of some materials in food packages.

Another complication arises when packages are subjected to sterilization processes using steam. The steam saturates the materials, altering the performance of humidity-sensitive materials during the time the package is drying.

The package designer must choose among about 20 different structural resins and a similar number of barrier resins. The number of possible combinations is, therefore, in the hundreds—too many to manually analyze.

The complexity of the package design problem makes it difficult to adequately describe a new barrier material through printed specifications. However, we found that a personal computer does have enough power to perform the necessary analysis.

The Development Process

Packaging Advisor was placed in full commercial operation nine months after development began. A rapid prototyping strategy was used to develop the system; we did not attempt to produce a full functional specification up front. The first prototype was demonstrated to potential users only three months after the start of the project, refinements were made over the next five months, and a final month was required for the duplication of diskettes and related materials.

The prototyping approach to system development had a number of advantages. At first, the complexity of the problem seemed a bit overwhelming. In addition to all the considerations described earlier, we discussed a number of other factors that could influence package design: type and location of handles, the need to stack some kinds of packages during transit and storage, and so on. The decision to start with the simplest plausible prototype helped us identify and focus on the truly important factors.

The prototypes helped the experts identify areas in which the system's knowledge was inaccurate or incomplete. The first prototype, for example, recommended combinations of materials that seemed implausible to one of the experts. After reflecting, he realized that the materials had different processing temperatures: One would vaporize before the other melted. Thus, the fact that we needed to include knowledge regarding processing temperatures became apparent.

The prototyping strategy also made it possible for the process of fielding the system to begin before the development process was complete. Prototype systems were shown to both customers and field sales representatives early in the development process. Several customers expressed strong interest based on the prototypes, giving the system a positive reputation with the field sales force before it was formally introduced. By the time the system was placed in production, customers were waiting to license it, and sales representatives were ready to communicate success stories to their peers. The early positive feedback from customers made it easier to maintain management's support for the project and helped build the developers' morale.

We estimate that at current prices, a system such as Packaging Advisor would cost approximately $50,000 to develop and field. This estimate includes costs for system development, documentation, and duplication services but does not include a charge for the services of domain experts.

Annual maintenance costs should also be considered. Because substantial enhancements to Packaging Advisor are planned, annual maintenance costs might approach development costs. In the case of Packaging Advisor, these costs are more than offset by license fees charged to customers and expenses for other marketing communications activities that have been displaced by the Packaging Advisor system.

The System

Packaging Advisor is a stand-alone system that runs on the IBM personal computer and compatibles. It consists of two components: the expert system front end, written in the Level5 shell from Information Builders, Inc., and a program written in the dBase III programming language and compiled with Clipper.

```
Packaging Advisor
Maximum Use Temperature
We need to know the maximum temperature the package will experi-
ence for a sustained period (over 2 minutes) during normal use. This
maximum temperature will normally be attained either during package
filling/sterilization or during heating of packages in microwave or con-
ventional ovens.
{We will rule out materials that cannot tolerate the indicated maximum
temperature. You may enter your temperature requirements directly if
you wish.}
    Room temperature or lower
    Pasteurization (71° C)
    Hot fill (85° C)
==> Retort sterilization and/or microwave oven heating (121° C)
    Heating in conventional ovens (230° C)
    Enter maximum usage temperature directly
```

Figure 1. Package Parameter Specification.

When the system is activated, the Level5 module assumes control and pursues the goal to "find the package specification desired by this user." Using backward chaining, the Level5 module fires appropriate rules and questions to determine the dimensions of the package, fabrication process to be used, desired shelf life, maximum allowable oxygen infiltration, desired optical properties, and a number of similar parameters. A typical question is shown in figure 1.

The Level5 module is designed to minimize the number of questions routinely asked. Inferencing is done to determine typical values for a number of secondary parameters. The parameters specified by the user and inferred by the system are then presented in the display shown in figure 2. In the example shown, the system has inferred that scrap will be recycled and that the scrap rate will be about 50 percent. The user can make any needed changes at this point in the interaction.

When the specification is complete, the Level5 module writes the parameters to a file and activates the Clipper module. This program searches two dBase files, which contain performance characteristics for structural and barrier resins, for suitable candidates. Each structural resin in the database is examined in turn. It can be ruled out for a number of reasons: incompatibility with the specified fabrication process, unsuitable optical properties, and so on.

```
                          Packaging Advisor
                     Package Requirements Summary

  Max. usage temp:              121° C.      Humidity inside package:    100%
  Storage temp:                  23° C.      Humidity outside pkg.:       60%
  Package area:           54.0 sq. inches    Retort sterilization:        Yes
  Package thickness:           30.0 mils
  Shelf life:        365 days
  Oxygen infiltration:           2.0 cc      Scrap recycled:              Yes
                                             Scrap rate:               50.0 %
  Processing method: Thermoforming
  Optical properties required: Opaque, translucent, or clear materials
  Location of barrier layer: centered       7.5 mils from Outside
  Must be covered by FDA food contact regulations:
          Structural resin: Yes
          Barrier resin: Yes
  Maximum thickness of barrier layer:       8.0 mils
  Minimum thickness of barrier layer:       0.4 mils

  Do you wish to make any changes in these parameters?

  ===> Accept these parameter values
        Change a parameter value
```

Figure 2. The Package Requirements Summary.

When a suitable structural resin is found, a check is performed to see whether a barrier resin is needed to meet the user's shelf life and oxygen-permeation requirements. If so, the barrier resin database is searched exhaustively for suitable candidates. A barrier resin for use in a food package must have suitable optical properties and regulatory agency status and a processing temperature range compatible with the structural resin and must provide an oxygen barrier more effective than that of the structural resin with which it is paired.

When a suitable pair of resins is located, calculations are performed to determine the required thickness of the barrier layer and the material costs for the package. A record is then written to a temporary results file. This file is sorted when all pairs of resins are processed, and a final output file is returned to the Level5 module for display to the user.

The system output is shown in figure 3. Alternative designs are ranked with the least costly first, as long as the user's shelf life requirement is satisfied. If no available materials have adequate oxygen barrier properties to meet this requirement, the package with the longest shelf life is shown first. The display in figure 3 is abbreviated; the actual system presents about 50 designs for the case shown.

Most of the barrier materials in the example are Du Pont products, but the system does not treat Du Pont materials preferentially. Cases exist where competitive materials are more cost effective, and the system presents these materials first. The decisions to include competitive materials and avoid preferential treatment of our own products reflect both faith in our product line and the desire to maximize the usefulness of the system to the customer. To further enhance the benefit of the system, it is designed to allow the customer to modify the databases to reflect personal resin costs and process economics.

The notes at the bottom of the screen in figure 3 provide additional information that could not easily be handled elsewhere in the system. We found, for example, that Food and Drug Administration regulations are complex. Some materials cannot be used with certain foods; others can be used with any food but only at certain temperatures; and so on. We added this type of information to footnotes rather than ask all the questions and code all the rules that would be necessary to cover these cases.

After analyzing a case, the user can elect to return to the package requirements summary screen shown in figure 2 and make whatever changes are desired for a what-if analysis. In this way, the user can gain a better understanding of the interrelationship between design criteria and package material costs. For example, the user might find that the cost difference between clear and translucent containers is greater than threefold.

How the System Was Used

The system was deployed on lap-top computers and used by field sales personnel, with assistance from headquarters staff, to introduce the new products to potential customers. The system was also made available for license to customers for their own use. A videotaped demonstration of the system was prepared so that field sales personnel who did not yet have lap-top computers or were uncomfortable using them could still demonstrate the system.

To our knowledge, Packaging Advisor is the first AI system designed to be the keystone of the marketing communications strategy for a new product line. Packaging Advisor is also a product in its own right—one that offers substantial benefits to customers. Because designers typically limit their analyses to a few favored materials, Packaging Advisor often suggests lower-cost alternatives than otherwise would have been considered. Often, the designer is also led to examine process changes involving variables such as scrap rate and package wall thickness. In fact, a number of customers have used the system to justify investments

Packages for Consideration

Structural Resin (Thickness, mils)	Barrier Resin (Thickness, mils)	Need BYNEL TIE	Mat'l Cost $U.S./M	Shelf Life (Days)	Notes
PP (29.0)	SELAR OH (30%) (1.0)	Y	34.129	365.0	B6
PP (27.3)	SELAR OH (44%) (2.7)	Y	50.210	365.0	B7
CPET (29.3)	SELAR OH (30%) (0.7)	Y	60.536	365.0	G3 S2 B6
PP talc filled (29.0)	SELAR OH (30%) (1.0)	Y	65.536	365.0	B6
PP (26.9)	PVDC (3.1)	Y	67.804	365.0	B5
CPET (28.0)	SELAR OH (44%) (2.0)	Y	70.175	365.0	G3 S2 B7
PP talc filled (27.3)	SELAR OH (44%) (2.7)	Y	77.804	365.0	B7
SELAR PT (29.1)	SELAR OH(30%) (0.9)	Y	79.068	365.0	G3 S20 B6

NOTES AND CAUTIONS
G2   Structure does not meet your shelf life requirement.
G3   Resin processing temperatures may not be compatible.

Structural Resin Notes
S2   CPET: not suitable for >50% alcoholic beverages; other limitations apply.
S18 Polysulfone: FDA regs specify only frozen/refrigerated storage.
S20 SELAR PT: cannot withstand significant internal pressure at retort temps.

Barrier Resin Notes
B4   NYLON MXD6: FDA restrictions apply; Check regulations.
B5   PVDC: Requires special fabrication eqpt; Barrier degrades at high temps.
B6   SELAR OH (30%): FDA regs specify max. 7 mil thickness & 100 Deg. C. Storage
B7   SELAR OH (44%): FDA regs specify max. 7 mil thickness & 100 Deg. C. Storage
B10 SELAR PA: FDA limited; not retortable in all cases.
B11 SELAR PT: cannot withstand significant internal pressure at retort temps.

Figure 3. System Output.

in process improvements. Clearly, the system is much more useful than printed product literature.

The Results

Goals for the system were threefold: (1) establish Du Pont as a technology leader in the eyes of barrier resin purchasers, (2) provide a means of demonstrating the value of Du Pont products, and (3) increase resin sales. We judge the system a success by all three criteria. The system received favorable reviews in the trade press and was well received by customers. The simultaneous introduction of technically advanced materials and Packaging Advisor established our position as the leading-edge supplier. Moreover, we are now selling enough resin to justify an expansion in production capacity.

Of course, it is hard to estimate what sales of the new materials would have been without Packaging Advisor. However, management believes that about 30 percent of resin sales are attributable to accounts with whom we made contact through Packaging Advisor. Without the system, we might never have been able to open the door at these accounts. The system enhanced the confidence of our sales representatives, enabled them to make more contacts, and improved the quality of their interactions with the customer.

Package designers often become deeply engrossed in their interactions with the system. On a number of occasions, we have seen them skip lunch or ask that a demonstration be extended so that they could complete their analysis. Few other marketing communications vehicles have been as successful at holding the attention of the target audience.

Conclusions

The Packaging Advisor case illustrates how expert systems technology can be used to provide a number of pragmatic benefits to a business. The project succeeded in codifying a significant body of technical knowledge and transforming this knowledge from a passive possession to a high-yielding asset. The resulting system gave us a significant competitive edge in the marketplace.

The use of an expert system tool and iterative prototyping methodology allowed us to compress the system development cycle and produce a system of significant technical complexity and strategic business value in less than a year's time.

A key factor in the success of the Packaging Advisor project was the strong sponsorship it received from the management of the business. The development of an expert system places substantial demands on a business unit. Experts are chronically short of time, and field sales staff can assimilate only a limited number of new programs. Controversial decisions, such as including data on competitor's products in Packaging Advisor, require expenditures of time and political capital. At Du Pont, we found that a business manager who is willing and able to muster the needed resources and successfully convert or override the inevitable naysayers is behind most of our success.

Although the economic and organizational costs of Packaging Advisor were significant, the payoff was high: The Du Pont company is building a barrier resins plant that almost surely could not have been justified without the competitive advantage provided by the system.

Index

Colophon

Editorial and Production Management by
The Live Oak Press, Palo Alto, California.

Copyedited by Elizabeth Ludvik.

Illustration by David Blatner.

Cover design by Spectra Media.

Composed in New Baskerville and Futura by
Parallax Productions on a Macintosh II
using Quark XPress and Adobe Illustrator88.

Output on a Linotronic 300 by National Colorite
Corp., New Berlin, Wisconsin.

Printed offset on 60 lb. Finch Opaque Smooth by
Malloy Lithographing, Inc., Ann Arbor, Michigan.